Methods in Molecular Biology

Series Editor
John M. Walker
School of Life and Medical Sciences
University of Hertfordshire
Hatfield, Hertfordshire, UK

For further volumes:
http://www.springer.com/series/7651

For over 35 years, biological scientists have come to rely on the research protocols and methodologies in the critically acclaimed *Methods in Molecular Biology* series. The series was the first to introduce the step-by-step protocols approach that has become the standard in all biomedical protocol publishing. Each protocol is provided in readily-reproducible step-by-step fashion, opening with an introductory overview, a list of the materials and reagents needed to complete. the experiment, and followed by a detailed procedure that is supported with a helpful notes section offering tips and tricks of the trade as well as troubleshooting advice. These hallmark features were introduced by series editor Dr. John Walker and constitute the key ingredient in,each and every volume of the *Methods in Molecular Biology* series. Tested and trusted, comprehensive and reliable, all protocols from the series are indexed in PubMed.

Plant Transcription Factors

Methods and Protocols

Edited by

José M. Franco-Zorrilla

Department of Plant Molecular Genetics, Centro Nacional de Biotecnología, Consejo Superior de Investigaciones Científicas (CSIC), Madrid, Madrid, Spain

Carmen Martín-Pizarro

Instituto de Hortofruticultura Subtropical y Mediterránea La Mayora, Universidad de Málaga-Consejo Superior de Investigaciones Científicas (IHSM, UMA-CSIC), Málaga, Málaga, Spain

Humana Press

Editors
José M. Franco-Zorrilla
Department of Plant Molecular Genetics
Centro Nacional de Biotecnología
Consejo Superior de Investigaciones Científicas (CSIC)
Madrid, Madrid, Spain

Carmen Martín-Pizarro
Instituto de Hortofruticultura Subtropical y Mediterránea La Mayora
Universidad de Málaga-Consejo Superior de Investigaciones Científicas (IHSM, UMA-CSIC)
Málaga, Málaga, Spain

ISSN 1064-3745 ISSN 1940-6029 (electronic)
Methods in Molecular Biology
ISBN 978-1-0716-4971-8 ISBN 978-1-0716-4972-5 (eBook)
https://doi.org/10.1007/978-1-0716-4972-5

This Humana imprint is published by the registered company Springer Science+Business Media, LLC, part of Springer Nature.
The registered company address is: 1 New York Plaza, New York, NY 10004, U.S.A.

Preface

Plant growth and adaptation to environmental conditions require strict control of gene activity, often involving transcriptional reprogramming. Transcription factors (TFs) are key regulators of gene expression, orchestrating precise spatiotemporal transcriptional responses to developmental signals and environmental stimuli in plants. As key nodes within complex gene regulatory networks, TFs can integrate endogenous signals with external stimuli, thereby ensuring plant growth under changing environmental conditions. Over the past decade, significant advances in experimental technologies and computational approaches have greatly enhanced our understanding of TFs and their binding sites across different tissues, developmental stages, and environmental scenarios. Importantly, applying these methodologies beyond model species, particularly to economically important crops, is crucial. This expansion enables the translation of fundamental discoveries into practical applications in plant breeding and biotechnology.

This book aims to provide a comprehensive guide covering the major aspects of plant transcription factors. The protocols described here will not be exclusively based on experimental setups, but also include analytical pipelines and examples to optimize information from databases and resources related to plant TFs. The book is organized into three parts. To enhance clarity and accessibility, we present a diverse set of methodological approaches, maintaining a balanced representation of experimental (wet-lab) and computational (dry-lab) techniques. In Part I, we present high-throughput methodologies for identifying TF-binding sites. This section includes standardized and optimized protocols for DAP-seq, ChIP-seq, and CUT&Tag, as well as ATAC-seq for chromatin accessibility profiling. These chapters emphasize the application of standardized protocols and computational tools in both model and crop species. Part II centers on the visualization and quantification of TF-DNA and TF-protein interactions. This section provides comprehensive methodologies to investigate the localization and timing of transcriptional complex formation in vivo. Furthermore, this part also covers methods to detect physical and functional interactions, crucial for linking TFs with their functional partners, and offering essential insights into their roles within complex regulatory networks. Part III includes several chapters containing integrative frameworks that combine transcriptomics with binding site prediction and gene regulatory network inference tools, essential for translating raw sequencing data into biological insight.

This volume is intended to serve as a comprehensive and practical resource for laboratories engaged in the study of plant transcription factors, thanks to the comprehensive, diverse, and high-quality content presented in the chapters.

Madrid, Madrid, Spain — *José M. Franco-Zorrilla*
Málaga, Málaga, Spain — *Carmen Martín-Pizarro*

Acknowledgments

We sincerely thank all the authors who generously contributed their valuable time and insights to this collection. We would also like to thank Dr. David Posé for his assistance.

Contents

Contributors

AMMAR ABDALRAHEM • *Estación Experimental de Aula Dei-CSIC, Zaragoza, Spain*

IÑIGO AGIRRE • *Institute for Integrative Systems Biology (I2SysBio), Universitat de València-CSIC, Paterna, Valencia, Spain*

EVA ÁLVAREZ • *Department of Plant Molecular Genetics, Centro Nacional de Biotecnología-CSIC, Madrid, Spain*

AGUSTÍN L. ARCE • *Instituto de Agrobiotecnología del Litoral, Universidad Nacional del Litoral– CONICET, Facultad de Bioquímica y Ciencias Biológicas, Centro Científico Tecnológico CONICET Santa Fe, Santa Fe, Argentina*

PATRICIA BALLESTER • *Instituto de Biología Molecular y Celular de Plantas, Consejo Superior de Investigaciones Científicas– Universitat Politècnica de València, Valencia, Spain*

ARACELI G. CASTILLO • *Plant-Virus Interaction Group, Instituto de Hortofruticultura Subtropical y Mediterránea La Mayora, Universidad de Málaga-Consejo Superior de Investigaciones Científicas (IHSM, UMA-CSIC), Málaga, Spain*

JAIME A. CASTRO-MONDRAGÓN • *Aix-Marseille Univ, INSERM UMR_S 1090, Theory and Approaches of Genome Complexity (TAGC), Marseille, France*

BRUNO CONTRERAS-MOREIRA • *Estación Experimental de Aula Dei-CSIC, Zaragoza, Spain*

THOMAS DEPUYDT • *Department of Plant Biotechnology and Bioinformatics, Ghent University, Ghent, Belgium; Center for Plant Systems Biology, VIB, Ghent, Belgium*

SHANSHAN DONG • *Fairy Lake Botanical Garden, Chinese Academy of Sciences, Shenzhen, China*

JOSÉ D. FERNÁNDEZ • *Centro de Genómica y Bioinformática, Universidad Mayor, Santiago, Chile; ANID-Millennium Science Initiative Program-Millennium Institute for Integrative Biology (iBio), Santiago, Chile*

CRISTINA FERRÁNDIZ • *Instituto de Biología Molecular y Celular de Plantas, Consejo Superior de Investigaciones Científicas– Universitat Politècnica de València, Valencia, Spain*

CARLOS R. FIGUEROA • *Laboratory of Plant Molecular Physiology, Institute of Biological Sciences, Campus Talca, Universidad de Talca, Talca, Chile; Millennium Nucleus for the Development of Super Adaptable Plants (MN-SAP), Santiago, Chile*

CHIARA FORESTI • *Department of Biotechnology, University of Verona, Verona, Italy*

JOSÉ M. FRANCO-ZORRILLA • *Department of Plant Molecular Genetics, Centro Nacional de Biotecnología, Consejo Superior de Investigaciones Científicas (CSIC), Madrid, Madrid, Spain*

YOLANDA GOGORCENA • *Estación Experimental de Aula Dei-CSIC, Zaragoza, Spain*

JOAQUÍN GRAU • *Department of Plant Molecular Genetics, Centro Nacional de Biotecnología-CSIC, Madrid, Spain*

SHAO-SHAN CAROL HUANG • *Center for Genomics & Systems Biology, New York University, New York, NY, USA*

NAJLA KSOURI • *Estación Experimental de Aula Dei-CSIC, Zaragoza, Spain*

TAMARA M. LEDESMA • *Instituto de Agrobiotecnología del Litoral, Universidad Nacional del Litoral– CONICET, Facultad de Bioquímica y Ciencias Biológicas, Centro Científico Tecnológico CONICET Santa Fe, Santa Fe, Argentina*

BINGBING LI • *College of Horticulture, China Agricultural University, Beijing, China*

LINZHOU LI • *State Key Laboratory of Agricultural Genomics, BGI-Shenzhen, Shenzhen, China*

MIAOMIAO LI • *Biosciences Division, Oak Ridge National Laboratory, Oak Ridge, TN, USA; Center for Bioenergy Innovation, Oak Ridge National Laboratory, Oak Ridge, TN, USA*

XIAOJING LI • *College of Horticulture, China Agricultural University, Beijing, China*

YANG LIU • *State Key Laboratory of Agricultural Genomics, BGI-Shenzhen, Shenzhen, China; Fairy Lake Botanical Garden, Chinese Academy of Sciences, Shenzhen, China*

ROSA LOZANO-DURÁN • *Department of Plant Biochemistry, Centre for Plant Molecular Biology (ZMBP), Eberhard Karls University, Tübingen, Germany*

SVITLANA LUKICHEVA • *Department of Plant Biotechnology and Bioinformatics, Ghent University, Ghent, Belgium; Center for Plant Systems Biology, VIB, Ghent, Belgium*

GABRIELLE MAGON • *Dipartimento di Agronomia Animali Alimenti Risorse Naturali e Ambiente (DAFNAE), University of Padova, Padova, Italy*

CARMEN MARTÍN-PIZARRO • *Instituto de Hortofruticultura Subtropical y Mediterránea La Mayora, Universidad de Málaga-Consejo Superior de Investigaciones Científicas (IHSM, UMA-CSIC), Málaga, Spain*

JOSÉ TOMÁS MATUS • *Institute for Integrative Systems Biology (I2SysBio), Universitat de València-CSIC, Paterna, Valencia, Spain*

FRANCESC MONTARDIT-TARDA • *Estación Experimental de Aula Dei-CSIC, Zaragoza, Spain*

PABLO MORALES-MARTÍNEZ • *Plant-Virus Interaction Group, Instituto de Hortofruticultura Subtropical y Mediterránea La Mayora, Universidad de Málaga-Consejo Superior de Investigaciones Científicas (IHSM, UMA-CSIC), Málaga, Spain*

JAVIER E. MORENO • *Instituto de Agrobiotecnología del Litoral, Universidad Nacional del Litoral – CONICET, Facultad de Bioquímica y Ciencias Biológicas, Centro Científico Tecnológico CONICET Santa Fe, Santa Fe, Argentina*

DAVID NAVARRO-PAYÁ • *Institute for Integrative Systems Biology (I2SysBio), Universitat de València-CSIC, Paterna, Valencia, Spain*

LUIS ORDUÑA • *Institute for Integrative Systems Biology (I2SysBio), Universitat de València-CSIC, Paterna, Valencia, Spain*

NICOLÁS MANOSALVA PÉREZ • *Department of Plant Biotechnology and Bioinformatics, Ghent University, Ghent, Belgium; Center for Plant Systems Biology, VIB, Ghent, Belgium*

MARÍA FLORENCIA PEROTTI • *Fruit Ripening and Quality Group, Instituto de Hortofruticultura Subtropical y Mediterránea La Mayora, Universidad de Málaga-Consejo Superior de Investigaciones Científicas (IHSM, UMA-CSIC), Málaga, Spain*

DAVID POSÉ • *Fruit Ripening and Quality Group, Instituto de Hortofruticultura Subtropical y Mediterránea La Mayora, Universidad de Málaga-Consejo Superior de Investigaciones Científicas (IHSM, UMA-CSIC), Málaga, Spain*

JULIA I. QÜESTA • *Centre for Research in Agricultural Genomics (CRAG), CSIC-IRTA-UAB, Barcelona, Spain*

ATILIO O. RAUSCH • *Instituto de Agrobiotecnología del Litoral, Universidad Nacional del Litoral – CONICET, Facultad de Bioquímica y Ciencias Biológicas, Centro Científico Tecnológico CONICET Santa Fe, Santa Fe, Argentina*

CARLOS SÁNCHEZ-GÓMEZ • *Fruit Ripening and Quality Group, Instituto de Hortofruticultura Subtropical y Mediterránea La Mayora, Universidad de Málaga-Consejo Superior de Investigaciones Científicas (IHSM, UMA-CSIC), Málaga, Spain*

ANTONIO SANTIAGO • *Institute for Integrative Systems Biology (I2SysBio), Universitat de València-CSIC, Paterna, Valencia, Spain*

Chaonan Shi • *Department of Plant Biochemistry, Centre for Plant Molecular Biology (ZMBP), Eberhard Karls University, Tübingen, Germany*

Jasper Staut • *Department of Plant Biotechnology and Bioinformatics, Ghent University, Ghent, Belgium; Center for Plant Systems Biology, VIB, Ghent, Belgium*

Huang Tan • *Department of Plant Biochemistry, Centre for Plant Molecular Biology (ZMBP), Eberhard Karls University, Tübingen, Germany*

Xiao-Yuan Tao • *Xianghu Laboratory, Hangzhou, China*

Meng-Bo Tian • *Center for Viticulture and Enology, College of Food Science and Nutritional Engineering, China Agricultural University, Beijing, China*

Juan Ignacio Tolopka • *Instituto de Agrobiotecnología del Litoral, Universidad Nacional del Litoral – CONICET, Facultad de Bioquímica y Ciencias Biológicas, Centro Científico Tecnológico CONICET Santa Fe, Santa Fe, Argentina*

Benjamin J. M. Tremblay • *Centre for Research in Agricultural Genomics (CRAG), CSIC-IRTA-UAB, Barcelona, Spain*

Aileen Turner • *Laboratory of Plant Molecular Physiology, Institute of Biological Sciences, Campus Talca, Universidad de Talca, Talca, Chile; Millennium Nucleus for the Development of Super Adaptable Plants (MN-SAP), Santiago, Chile*

Jacques van Helden • *Aix-Marseille Univ, INSERM UMR_S 1090, Theory and Approaches of Genome Complexity (TAGC), Marseille, France*

Klaas Vandepoele • *Department of Plant Biotechnology and Bioinformatics, Ghent University, Ghent, Belgium; Center for Plant Systems Biology, VIB, Ghent, Belgium; VIB Center for AI & Computational Biology, VIB, Ghent, Belgium*

Alessandro Vannozzi • *Dipartimento di Agronomia Animali Alimenti Risorse Naturali e Ambiente (DAFNAE), University of Padova, Padova, Italy*

Alvaro Vidal • *Institute for Integrative Systems Biology (I2SysBio), Universitat de València-CSIC, Paterna, Valencia, Spain*

Elena A. Vidal • *Centro de Genómica y Bioinformática, Universidad Mayor, Santiago, Chile; ANID-Millennium Science Initiative Program-Millennium Institute for Integrative Biology (iBio), Santiago, Chile*

Jia-Wei Wang • *National Key Laboratory of Plant Molecular Genetics (NKLPMG), CAS Center for Excellence in Molecular Plant Sciences (CEMPS), Institute of Plant Physiology and Ecology (SIPPE), Chinese Academy of Sciences (CAS), Shanghai, China; School of Life Science and Technology, ShanghaiTech University, Shanghai, China; Key Laboratory of Plant Carbon Capture, CAS, Shanghai, China; New Cornerstone Science Laboratory, Shanghai, China*

Sibo Wang • *State Key Laboratory of Agricultural Genomics, BGI-Shenzhen, Shenzhen, China*

Sara Zenoni • *Department of Biotechnology, University of Verona, Verona, Italy*

Chen Zhang • *Institute for Integrative Systems Biology (I2SysBio), Universitat de València-CSIC, Paterna, Valencia, Spain*

Paz E. Zuñiga • *Laboratory of Plant Molecular Physiology, Institute of Biological Sciences, Campus Talca, Universidad de Talca, Talca, Chile; Millennium Nucleus for the Development of Super Adaptable Plants (MN-SAP), Santiago, Chile; Doctoral Program in Sciences Mention in Plant Biology and Biotechnology, Institute of Biological Sciences, Campus Talca, Universidad de Talca, Talca, Chile*

Part I

High-Throughput Experimental Identification of Binding Sites

Chapter 1

Standardized Protocols and Bioinformatic Pipelines for Conducting DAP-seq Experiments in Non-model Plant Crops

Chen Zhang, Chiara Foresti, Luis Orduña, Miaomiao Li, Gabrielle Magon, Meng-Bo Tian, Alvaro Vidal, Antonio Santiago, David Navarro-Payá, Iñigo Agirre, Alessandro Vannozzi, Sara Zenoni, Shao-shan Carol Huang, and José Tomás Matus

Abstract

The cistrome comprises genomic loci that regulate gene expression, playing a crucial role in defining cellular identity and function. Analyzing cistrome data reveals key molecular mechanisms underlying grapevine growth, development, and environmental responses. Identifying transcription factors that bind specific DNA sequences allows researchers to dissect the complex regulatory networks controlling gene expression. Moreover, this analysis can help pinpoint targets for crop improvement, as traits like fruit quality, disease resistance, and abiotic stress tolerance are often regulated by transcription factors. DNA affinity purification sequencing (DAP-seq) is a high-throughput, cost-effective method for mapping the cistrome, providing valuable insights into transcriptional regulation. This technique relies on the in vitro affinity purification of genomic DNA-protein complexes, followed by high-throughput sequencing of eluted DNA fragments. Unlike other in vitro DNA-binding assays, such as protein-binding microarrays (PBM) and systematic evolution of ligands by exponential enrichment (SELEX), DAP-seq allows transcription factors to interact directly with plant-derived genomic DNA, capturing all potential binding sites. The resulting data closely resemble those from chromatin immunoprecipitation sequencing (ChIP-seq) but are obtained much faster. Initially developed in *Arabidopsis*, DAP-seq has since been applied to several crops, including maize, tomato, and grapevine, generating extensive cistrome datasets and deepening our understanding of gene regulatory regions. However, despite its power in elucidating crop biology, DAP-seq faces certain limitations, particularly concerning the size and complexity of plant genomes. This chapter presents detailed protocols for DAP-seq studies aimed at the unbiased identification of transcription factor binding sites in crops. Additionally, we outline a standardized pipeline for DAP-seq data analysis, encompassing raw sequencing data processing (i.e., trimming, filtering, and read alignment), as well as peak calling and motif discovery analysis. This approach enables the efficient and scalable identification of transcription factor binding profiles in diverse crop species.

First authorship: Chen Zhang and Chiara Foresti

Shared corresponding authorship: Sara Zenoni, Shao-shan Carol Huang and José Tomás Matus

José M. Franco-Zorrilla and Carmen Martín-Pizarro (eds.), *Plant Transcription Factors: Methods and Protocols*,
Methods in Molecular Biology, vol. 2985, https://doi.org/10.1007/978-1-0716-4972-5_1,

Key words Cistrome, DAP-seq, Transcription factors, Gene regulation, Crops

1 Introduction

Gene expression regulation is a dynamic and multilayered process that determines transcript abundance in a cell. Among its various mechanisms, transcriptional regulation plays a central role, involving transcription factors (TFs) that interact with specific DNA motifs to modulate gene expression. Understanding TF function is crucial for agriculture, as it guides biotechnological strategies aimed at improving crop quality and resilience to environmental stress. Traditionally, TF studies followed a reductionist approach, examining individual factors in isolation. However, the complexity of transcriptional regulatory networks, where TFs act in concert as activators or repressors, has led to a shift toward systems biology. The rapid expansion of next-generation sequencing (NGS) has fueled this transition, enabling the study of regulatory genomics in crops, where transcriptional control influences phenotypic plasticity and fruit quality.

The diversity of plant transcriptional regulation arises from the large number of TFs encoded in plant genomes, which have expanded through gene duplication and diversification. Evolutionary studies show that TF family expansion correlates with organismal complexity and has facilitated the emergence of novel adaptive traits [1]. Despite this, studying TF function in non-model plants remains challenging due to limited genetic resources, such as sequenced and annotated genomes and mutant collections. To overcome these limitations, in silico and in vitro approaches can be employed to predict TF function and define their cistromes.

Several high-throughput methods exist for mapping TF binding sites at a genomic scale, including ChIP-seq, CUT&Tag, DamID, and SELEX-seq. ChIP-seq remains widely used in both model and non-model species, contributing to databases such as PlantTFDB [2] and PlantPan [3]. However, its effectiveness relies on antibody specificity and sensitivity, making it impractical for many TFs and plant species [4]. Although transgenic lines expressing epitope-tagged TFs offer a solution, generating such plants is labor-intensive and time-consuming, especially in non-model crops. DNA affinity purification sequencing (DAP-seq) was developed as an in vitro alternative to ChIP-seq, enabling genome-wide TF binding site identification without requiring antibodies or transgenic lines [5, 6]. In DAP-seq, genomic DNA is extracted, fragmented, and used to construct a sequencing library. The TF of interest, expressed in vitro with an affinity tag, is incubated with this DNA, allowing binding sites to be captured and sequenced. DAP-seq provides high-resolution TF binding profiles, and while it captures a large fraction of ChIP-seq peaks, its in vitro nature means some interactions requiring cofactors may be missed.

First introduced in *Arabidopsis*, DAP-seq has been subsequently applied to various crops and fruit, such as maize [7], grapevine [8], tomato [9], melon [10], and banana [11], producing extensive cistrome datasets and enhancing our insight in gene regulation. DAP-seq analysis is affected by the size and complexity of the species' genome. Large, repetitive sequences and high heterozygosity can pose challenges for accurate binding site identification. This chapter offers a detailed guide for performing DAP-seq in non-model crops, covering experimental protocols and computational analysis. The provided analysis pipeline, which includes data processing, peak calling, and motif enrichment analysis, is designed for scalability and ease of use across diverse plant species.

2 Materials

2.1 Genomic DNA Extraction (Adapted from [6])

1. Mortar and pestle.
2. Liquid nitrogen.
3. Tubes (50 mL, 1.5 mL, and thin-wall clear 0.5 mL).
4. Vortex.
5. Centrifuge.
6. Water bath.
7. Chemical hood.
8. Qubit™ fluorometer (ThermoFisher Scientific).
9. NanoDrop One and One^c Spectrophotometer (ThermoFisher Scientific).
10. Crude Nuclei buffer, CNB, pH 7.6 (0.25 M NaCl, 0.2 M Tris-HCl, pH 7.6, 2.5% PVP40, 0.05 M Na_2EDTA, pH 8, 0.1% β -ME).
11. EtOH/Sark buffer (0.5 M NaCl, 0.2 M Tris-HCl, pH 8, 2.5% PVP 40.000, 0.05 M Na_2EDTA, pH 8, 3% N-lauroyl-sarcosine, 20% ethanol, 1% β-ME).
12. Chloroform.
13. Isopropanol.
14. 70% ethanol.
15. TE buffer (10 mM Tris-HCl, pH 8, 1 mM Na_2EDTA, pH 8).
16. RNase (20 mg/mL).
17. 7.5 M NaH_4OAc.
18. Qubit™ dsDNA BR assay kit (ThermoFisher Scientific).

2.2 Genomic DNA Purification (Optional)

1. Tubes (1.5 mL and thin-wall clear 0.5 mL).
2. Vortex.
3. Centrifuge.
4. Chemical hood.
5. Qubit™ fluorometer (ThermoFisher Scientific).
6. NanoDrop One and One^{c} Spectrophotometer (ThermoFisher Scientific).
7. Chloroform: isoamyl alcohol (24:1).
8. Absolute ethanol.
9. 70% ethanol
10. TE buffer (10 mM Tris-HCl, pH 8, 1 mM Na_2EDTA, pH 8).
11. Qubit™ dsDNA BR assay kit (ThermoFisher Scientific).

2.3 TruSeq Illumina gDNA Library Preparation (Adapted from [6])

1. Covaris S2 focused sonicator.
2. microTUBEs AFA Fiber Pre-Slit Snap-Cap 6 × 16 mm tube (Covaris).
3. Tubes (low-bind 1.5 mL and thin-wall clear 0.5 mL).
4. Vortex.
5. Centrifuge.
6. Dynamag™-2, Beads separation (Invitrogen).
7. Qubit™ fluorometer (ThermoFisher Scientific).
8. End-It kit (Lucigen).
9. QIAquick PCR Purification Kit (Qiagen).
10. Klenow 3′-5′ exo- (NEB).
11. NEBuffer2 (NEB).
12. dATP 100 mM (NEB).
13. T4 DNA Ligase 400,000 U/mL (NEB).
14. 30 μM annealed Y-Adapter: Each adapter must be diluted to 100 μM with sterile water. Combine 150 μL of each adapter with 200 μL of sterile water to reach the total of 500 μL. Incubate the adapter mixture at 96 °C for 2 min, and then cool to room temperature (22 °C). Store the annealed Y-Adapter at −20 °C.
15. EB buffer (Elution Buffer, Qiagen).
16. AMPure X beads (Beckman Coulter).
17. 80% ethanol.
18. 3 M NaOAc, pH 5.5 (Invitrogen).
19. Qubit™ dsDNA BR assay kit (ThermoFisher Scientific).
20. KAPA Library Quantification kit (Illumina) or Bioanalyzer High Sensitivity DNA (Agilent, optional).

2.4 TruSeq Illumina gDNA Libraries Amplification (Optional)

1. Include all materials mentioned in Subheading 2.3.
2. 0.2 mL PCR tubes.
3. Phusion High-Fidelity DNA Polymerase (NEB).
4. Illumina i5 and i7 indexes.

2.5 Transcription Factor In Vitro Expression

1. TF-tagged plasmid (e.g., pIX-Halo-ORF, GST-Halo-ORF, or others with Halo tag).
2. TNT Coupled Reticulocyte Lysate Systems (Promega).
3. RNasin Ribonuclease Inhibitor (Promega).
4. 96 Well Non-Skirted PCR Plate (Azenta).
5. EB buffer (Elution Buffer, Qiagen).

2.6 Dap

1. Dynamag™-2, Beads separation (Invitrogen).
2. DynaMag™-96 Side Skirted PCR Magnet (Invitrogen).
3. Tube revolver (ThermoFisher).
4. Tubes (0.2 mL, low bind 1.5 mL, and 96 Well Non-Skirted PCR Plate).
5. Magne Halo-tag beads (Promega).
6. PBS buffer: Dissolve 8 g of NaCl, 0.2 g of KCl, 1.44 g Na_2HPO_4, and 0.24 g of KH_2PO_4 in 800 mL of sterile water. Adjust the pH to 7.4 with HCl. Add water to reach 1 L and sterilize by autoclaving.
7. NP40.
8. PBS + NP40 buffer (1×PBS, 0.005%NP40).
9. EB buffer (Elution Buffer, Qiagen).

2.7 PCR Enrichment

1. Tubes (0.2 mL and low bind 1.5 mL).
2. Phusion High-Fidelity DNA Polymerase (NEB).
3. Illumina i5 and i7 indexes.
4. Dynamag™-2, Beads separation (Invitrogen).
5. AMPure X beads (Beckman Coulter).
6. EB buffer (Elution Buffer, Qiagen).
7. Qubit™ dsDNA HS assay kit (ThermoFisher Scientific).
8. KAPA Library Quantification kit (Illumina) or Bioanalyzer High Sensitivity DNA (Agilent).

2.8 Computational Requirements for Bioinformatic Analysis

A bioinformatic analysis starting from DAP-seq raw reads requires the following software and packages (the versions are indicated for reproducibility; more recent versions should also work):

- python3 (preinstalled in Linux based distributions) (v3.11.4).
- R (v4.2.1) and the following R packages:

  - ChIPpeakAnno (v3.32.0) [12].
  - rtracklayer v1.58.0.
- biotools (v2).
- bowtie2 (v2.4.5) [13].
- samtools (v1.9).
- bamCoverage (v3.5.2).
- fastp (v0.21.0).
- GEM (v3.4).
- FastQC (v0.11.9).

2.9 Adapter and Index Sequences

Oligonucleotide sequences are indicated in Table 1.

3 Methods

3.1 Genomic DNA Extraction (Adapted from [14]; See *Note 1)*

1. Put 1 g of ground tissue in a 50 mL Falcon tube.
2. Add 12.5 mL of CNB (with β-ME added just before) and vortex the mixture.
3. Centrifuge at 3000×*g* for 10 min at 4 °C and discard the supernatant.
4. Add 5 mL of EtOH/Sark buffer (with β-ME and EtOH added just before) and resuspend the pellet by vortexing (*see* **Note 2**).
5. Put the 50 mL Falcon tube at 55 °C for 30 min with frequent shaking.
6. Add 5 mL of chloroform and vortex.
7. Centrifuge at 3000×*g* for 20 min at 20 °C.
8. Add 2.5 mL of isopropanol into a new 50 mL Falcon tube.
9. Transfer the upper phase of the 50 mL Falcon tube into the 50 mL Falcon tube with isopropanol, prepared before.
10. Mix very well by inverting the Falcon tube until a white precipitate is visible.
11. Leave at room temperature (RT, 22 °C) for 15 min.
12. Centrifuge at 3000×*g* for 15 min at 20 °C and discard the supernatant.
13. Put the 50 mL Falcon tube upside down on paper layers to better discard the remaining drops.
14. Add 5 mL of 70% EtOH and wash the pellet, vortexing and centrifuging at 3000×*g* for 20 min at 20 °C, and discard the supernatant. Repeat this step.

Table 1
Adapter and index oligonucleotides used in this protocol

Name	Sequences	Purification
DAP-adapter-1	ACACTCTTTCCCTACACGACGCTCTTCCGATCT	HPLC
DAP-adapter-2	/5Phos/GATCGGAAGAGCACACGTCTGAACTCCAGTCAC	HPLC
iTruS_i5_D501	AATGATACGGCGACCACCGAGATCTACACTATAGCCTACACTC TTTCCCTAC	Standard
iTruS_i5_D502	AATGATACGGCGACCACCGAGATCTACACATAGAGGCACACTC TTTCCCTAC	Standard
iTruS_i5_D503	AATGATACGGCGACCACCGAGATCTACACCCTATCCTACACTC TTTCCCTAC	Standard
iTruS_i5_D504	AATGATACGGCGACCACCGAGATCTACACGGCTCTGAACAC TCTTTCCCTAC	Standard
iTruS_i5_D505	AATGATACGGCGACCACCGAGATCTACACAGGCGAAGACAC TCTTTCCCTAC	Standard
iTruS_i5_D506	AATGATACGGCGACCACCGAGATCTACACTAATCTTAACACTC TTTCCCTAC	Standard
iTruS_i5_D507	AATGATACGGCGACCACCGAGATCTACACCAGGACGTACAC TCTTTCCCTAC	Standard
iTruS_i5_D508	AATGATACGGCGACCACCGAGATCTACACGTACTGACACACTC TTTCCCTAC	Standard
iTruS_i7_D701	CAAGCAGAAGACGGCATACGAGATCGAGTAATGTGACTGGAG TTCAG	Standard
iTruS_i7_D702	CAAGCAGAAGACGGCATACGAGATTCTCCGGAGTGACTGGAG TTCAG	Standard
iTruS_i7_D703	CAAGCAGAAGACGGCATACGAGATAATGAGCGGTGACTGGAG TTCAG	Standard
iTruS_i7_D704	CAAGCAGAAGACGGCATACGAGATGGAATCTCGTGACTGGAG TTCAG	Standard
iTruS_i7_D705	CAAGCAGAAGACGGCATACGAGATTTCTGAATGTGACTGGAG TTCAG	Standard
iTruS_i7_D706	CAAGCAGAAGACGGCATACGAGATACGAATTCGTGACTGGAG TTCAG	Standard
iTruS_i7_D707	CAAGCAGAAGACGGCATACGAGATAGCTTCAGGTGACTGGAG TTCAG	Standard
iTruS_i7_D708	CAAGCAGAAGACGGCATACGAGATGCGCATTAGTGACTGGAG TTCAG	Standard
iTruS_i7_D709	CAAGCAGAAGACGGCATACGAGATCATAGCCGGTGACTGGAG TTCAG	Standard
iTruS_i7_D710	CAAGCAGAAGACGGCATACGAGATTTCGCGGAGTGACTGGAG TTCAG	Standard

(continued)

Table 1
(continued)

Name	Sequences	Purification
iTruS_i7_D711	CAAGCAGAAGACGGCATACGAGATGCGCGAGAGTGACTGGAG TTCAG	Standard
iTruS_i7_D712	CAAGCAGAAGACGGCATACGAGATCTATCGCTGTGACTGGAG TTCAG	Standard

15. Put the tube upside down on some paper layers to better discard the supernatant drops.
16. Let the pellet air dry.
17. Resuspend the pellet with 600 μL of TE buffer, adding with 3 μL of RNase.
18. Place the tube at 4 °C overnight.
19. Transfer the gDNA solution to a 1.5 mL tube.
20. Add 300 μL of 7.5 M NH_4OAc.
21. Centrifuge at 13,000 rpm for 10 min at RT.
22. Transfer the supernatant to a new 1.5 mL tube.
23. Add 300 μL of isopropanol.
24. Mix by inverting until a precipitate is seen.
25. Centrifuge at 14,000×*g* for 10 min at RT and discard the supernatant.
26. Wash the pellet with 1 mL of 70% EtOH twice by adding the EtOH, vortexing and centrifuging at 13,000 rpm for 10 min at RT and discard the supernatant.
27. Let the pellet air dry.
28. Resuspend the pellet with 300 μL of TE buffer (*see* **Note 3**).
29. Leave the tube at 4 °C overnight for a better resuspension.
30. Measure the gDNA concentration using the Qubit™ (*see* **Note 4**) and the 260/230 and 260/280 ratio using the NanoDrop (*see* **Note 5**).

3.2 Genomic DNA Purification (Optional; See Note 6)

1. Add an equal volume of chloroform—isoamyl alcohol (24:1)—and vortex.
2. Centrifuge at 14,000×*g* for 10 min at RT.
3. Transfer the supernatant to a new 1.5 mL tube.
4. Add an equal volume of chloroform—isoamyl alcohol (24:1)—and vortex.
5. Centrifuge at 14,000×*g* for 10 min at RT.
6. Transfer the supernatant to a new 1.5 mL tube.

7. Add 2.5 volumes of absolute ethanol and mix very well by inverting until a white precipitate is seen.
8. Keep it at RT for 15 min.
9. Centrifuge at 14,000×*g* rpm for 10 min at RT and discard the supernatant.
10. Wash the pellet with 1 mL of 70% EtOH and vortex.
11. Centrifuge for 10 min at RT and discard the supernatant.
12. Let the pellet air dry.
13. Resuspend the pellet with 100 μL of TE buffer.
14. Leave the tube at 4 °C overnight for a better resuspension.
15. Measure the gDNA concentration using the Qubit™ and the 260/230 and 260/280 ratio using the NanoDrop.

3.3 TruSeq Illumina gDNA Library Preparation (Adapted from [6])

1. Set up Covaris S2 with chill and degas water.
2. Dilute 7.5 μg of gDNA using EB buffer to a final volume of 130 μL, incubate 5 min at RT, and load it into a Snap-cap Covaris tube (*see* **Note 7**).
3. Run the 200 bp fragmentation program on the Covaris S2:
   Mode: frequency sweeping
   - Duty cycle: 10%.
   - Intensity: 5.
   - Cycles/burst: 200.
   - Time: 60 sec.
   - Number of cycles: 3.
   - Water level: 12.
   - T = <10 °C.
4. Immediately transfer the sheared sample (around 125 μL) to a 1.5 mL low-bind tube.
5. Add AMPure XP beads (1×) to the sheared samples and vortex. Mix the beads thoroughly in advance by vortexing.
6. Quick spin and incubate at RT for 10 min.
7. Place on the magnetic rack for 5 min.
8. Remove most of the supernatant with the pipette (*see* **Note 8**).
9. Immediately add 500 μL 80% EtOH.
10. Remove and rotate tube 180 degrees on the magnet six times to wash the beads (*see* **Note 9**).
11. Wait 1 min and pipette off the EtOH.
12. Repeat the wash (**steps 9–11**).
13. Spin and pipette off all the remaining EtOH.

14. Dry tubes with caps open at 37 °C on the magnet until cracks form in beads (*see* **Note 10**).
15. Resuspend beads in 35 μL of EB buffer and mix well by pipetting (*see* **Note 11**).
16. Incubate at RT for 10 min.
17. Place the tubes on the magnet for 3 min and then aspirate supernatant into a new tube, being careful not to transfer beads (*see* **Note 12**).
18. Perform the end repair reaction (e.g., End-It kit; Lucigen), adding 5 μL of 10× buffer, 5 μL of dNTP mix, 5 μL of ATP, and 1 μL of End-It Enzyme mix to the samples. Other kits can also be used (*see* **Note 13**).
19. Incubate at RT for 45 min.
20. Purify using the Qiaquick PCR purification kit (*see* **Note 14**).
21. Elute with 33 μL of EB buffer.
22. Perform the A-Tail reaction, adding 5 μL of NEBuffer2, 10 μL of 1 mM dATP, and 3 μL of Klenow (3′-5′ exo-) to the samples (*see* **Note 15**).
23. Incubate at 37 ° C for 30 min.
24. Purify using the Qiaquick PCR purification kit.
25. Elute with 32 μL of EB buffer.
26. Perform the Adapter Ligation reaction, adding 5 μL of 10× Ligase buffer, 10 μL of 30 μM Y-adapter, and 5 μL of T4 DNA Ligase (*see* **Note 16**).
27. Incubate at 16 °C overnight.
28. Heat-inactivate the reaction at 70 °C for 10 min and put in ice.
29. Perform the AMPure XP beads cleaning (**steps 5–17**), resuspending the beads in 32 μL of EB buffer.
30. Check the concentration using the Qubit or fragment analysis.
31. Check library quality using the bioanalyzer or the KAPA Library Quantification Kit (optional).

3.4 TruSeq Illumina gDNA Libraries Amplification for ampDAP-Seq (Optional; Adapted from [6]; See Note 17)

1. Follow all steps mentioned in Subheading 3.3 and obtain ligated DNA fragments.
2. Prepare the PCR reactions by adding 10 μL of 5× HF buffer, 2.5 μL of 10 mM dNTPs, 1 μL of 25 μM i5 primer, 1 μL of 25 μM i7 primer, 1 μL of Phusion polymerase, 15 ng of ligated DNA fragments, and water to reach 50 μL of PCR reaction samples.
3. Set up the PCR:

   95 °C × 2 min

   98 °C × 30 s.

11 cycles of (98 °C × 15 s; 60 °C × 30 s; 72 °C × 2 min)
72 °C × 10 min
4 °C HOLD.

4. Perform the AMPure XP beads cleaning (**steps 5–17** from Subheading 3.3), resuspending the beads in 35 μL of EB buffer.
5. Check the concentration using the Qubit or fragment analysis.
6. Check library quality using the bioanalyzer or the KAPA Library Quantification Kit (optional).

3.5 Transcription Factor In Vitro Expression

1. Prepare the TNT Coupled Reticulocyte Lysate reaction, adding (to the 200 μL lysate containing tube) 16 μL of TNT Reaction buffer, 8 μL of Minus Met 1 mM, 8 μL of Minus Leu 1 mM, 8 μL of RNasin Ribonuclease Inhibitor, 8 μL of TNT RNA Polymerase SP6, and 72 μL of water (*see* **Note 18**).
2. Aliquot 39 μL of the mix to eight new 0.2 mL PCR tubes (*see* **Note 19**).
3. Prepare the pIX-HALO-ORF for a total of 1 μg of plasmid in 10 μL of EB buffer.
4. Add the 10 μL of EB diluted pIX-Halo-ORF to each eight TNT Coupled Reticulocyte Lysate reaction tube, and pipette slowly up and down to mix.
5. Quick spin and incubate at 30 °C for 2 h (*see* **Note 20**).
6. Check the Halo-ORF clones with immunoblotting, using 5 μL of the in vitro expressed proteins and an anti-Halo monoclonal antibody.

3.6 DAP

1. Put 10 μL of Magne Halo-tag beads per sample in a 1.5 mL tube (*see* **Note 18**). This would be 80 μL of Magne Halo-tag beads for a total of eight samples (in a 1.5 mL tube).
2. Add 1 mL of PBS + NP40 buffer to the Magne Halo-tag beads.
3. Flick the tube and place it on the magnet for 1 min.
4. Remove the supernatant.
5. Repeat **steps 2–4** another two times.
6. Resuspend the beads with 360 μL of PBS + NP40 buffer.
7. Add 45 μL of washed Magne Halo-tag beads to each of the 45 μL of the remaining in vitro expressed protein reaction (*see* **Note 21**).
8. Rotate at RT for 1 h.
9. Place the PCR tubes on the magnet for 1 min.
10. Remove the supernatant and wash with 90 μL of PBS + NP40 buffer, flicking gently the tube to mix.

11. Let the beads settle.
12. Repeat **steps 9–11** (three times).
13. Place the PCR tubes on the magnet for 1 min and remove the supernatant.
14. Finally, resuspend the beads with 40 μL of PBS + NP40 buffer.
15. Dilute 100–1000 ng of adapter ligated gDNA libraries in EB buffer to a final volume of 40 μL (*see* **Note 22**).
16. Add the 40 μL of EB diluted gDNA libraries to the 40 μL of beads (each sample will have a total final volume of 80 μL) (*see* **Note 21**).
17. Rotate at RT for 1–2 h.
18. Place the PCR tubes on the magnet for 1 min and remove the supernatant.
19. Wash the beads with 90 μL of PBS + NP40 buffer for six times (Subheading 3.6, **steps 10–13**).
20. Resuspend the beads with 25 μL of EB buffer and transfer to a new PCR tube.
21. Heat tubes at 98 °C for 10 min (*see* **Note 23**).
22. Place the tubes on ice for 5 min.
23. Place the tubes on the magnet and transfer the supernatant to new PCR tubes (*see* **Note 24**).

3.7 PCR Enrichment (See Note 25)

1. Prepare the PCR enrichment reaction by adding 10 μL of 5× HF buffer, 2.5 μL of 10 mM dNTPs, 1 μL of 25 μM i5 primer, 1 μL of 25 μM i7 primer, 1 μL of Phusion polymerase, and 9.5 μL of water to each 25 μL of the DAP obtained sample.
2. Set up the PCR:

   95 °C × 2 min

   98 °C × 30 s.

   20 cycles of (98 °C × 15 s; 60 °C × 30 s; 72 °C × 1 min)

   72 °C × 10 min

   4 °C × HOLD.
3. Transfer the PCR reactions to 1.5 mL tubes.
4. Perform the AMPure XP beads cleaning (**steps 5–17** from Subheading 3.3), resuspending the beads in 22 μL of EB buffer.
5. Check the concentration using the Qubit.
6. Check library quality using the bioanalyzer or the KAPA Library Quantification Kit to confirm the absence of adapter dimers prior to sequencing (*see* **Note 26**).
7. Store the samples at −20 °C.

3.8 Sequencing and Downstream Bioinformatics Analysis

1. Sequence at least two independent technical replicates per TF and input (*see* **Note 27**).
2. Sequencing of libraries can be set at 5–20 million and 1 × 75bp single-end or 2 × 75 paired-end reads (*see* **Note 28**).
3. Trimming of the libraries using fastp software. For single-end libraries, use the following parameters: "-w 16 --n_base_limit 5 cut_front_window_size 1 cut_front_mean_quality 30 --cut_-front cut_tail_window_size 1 cut_tail_mean_quality 30 --cut_-tail -l 20." For paired-end libraries, use the following parameters: "-w 16 --detect_adapter_for_pe --n_base_limit 5 cut_front_window_size 1 cut_front_mean_quality 30 --cut_-front cut_tail_window_size 1 cut_tail_mean_quality 30 --cut_-tail -l 20."
4. Alignment of the trimmed libraries using bowtie2 with default parameters (*see* **Note 29**).
5. Sorting of alignments according to chromosomal coordinates and removal of multiple aligned reads using Samtools.
6. Peak calling comparing TF libraries against input (pHALO/GST) libraries using GEM peak caller with the following parameters: "--t 1 --q 2 --f BAM --outNP --outBED --outJASPAR --outMEME --outHOMER --k_min 6 --k_max 20 --k_seqs 600 --k_neg_dinu_shuffle" (*see* **Notes 30** and **31**).
7. Assignment of peaks to their corresponding gene using ChIP-peakAnno R package with default parameters (i.e., NearestLocation).
8. Generation of the DNA-binding motif obtained from the enrichment analysis of the top 600 most significant peaks sequences with MEME suite using default parameters (*see* **Note 32**).
9. For identification of regulated genes, an overlap between the DAP-seq bound genes and any transcriptomic dataset (i.e., an overexpression of the analyzed TF) is highly recommended.

4 Notes

1. DAP-seq is highly dependent on gDNA quality, making it crucial to establish an effective nucleic acid extraction methodology to maximize efficiency in downstream applications. Various protocols have been tested and compared based on yield and purity. The protocol described by [14], with minor adaptations, has proven effective for young leaves of grapevine [15], sorghum [16], and tomato (unpublished results), as well as for leaves and flowers of cannabis (unpublished results). However, DNA yield decreases with senescence progression. A typical

extraction should yield approximately 40–70 μg of gDNA per gram of ground tissue.

2. After adding the EtOH/Sark buffer, incubating at 55 °C for 30 min with constant shaking at 100–150 rpm has proven effective in processing leaf tissues that are particularly coriaceous and rich in polyphenols and polysaccharides, such as grapevine leaves.
3. Low amount of gDNA is extracted from green berries, but the DNA is still highly pure, so multiple extractions are needed for fruit tissues. For these cases, it is better to resuspend the pellet of three independent extractions in 300 μL TE buffer.
4. Since the Qubit™ system relies on fluorescence emitted by a dye that selectively binds to a specific analyte, it provides more accurate quantification than the NanoDrop. In contrast, NanoDrop measurements can be easily compromised by contaminants that absorb at the same wavelength as nucleic acids, leading to potential overestimation or inaccuracies.
5. A 260/280 ratio of ~1.8 is generally accepted as "pure" for DNA. If the ratio is appreciably lower, it may indicate the presence of protein, phenol, or other contaminants that absorb strongly at or near 280 nm. This 260/230 ratio is used as a secondary measure of nucleic acid purity. The 260/230 values for "pure" nucleic acid are often higher than the respective 260/280 values. Expected 260/230 values are commonly in the range of 2.0–2.2. If the ratio is appreciably lower than expected, it may indicate the presence of contaminants which absorb at 230 nm.
6. This step is needed if the results of gDNA extraction do not reach the minimum values of 260/230 and 260/280 absorption. All the steps listed here are made as an extra control, an extra purification step, or handmade reagents. Also, due to the abundance in polyphenols and polysaccharides, there is the possibility to perform an additional purification with phenol/chloroform/isoamyl alcohol (24:24:1) before the chloroform/isoamyl alcohol (24:1) reported in [14].
7. Best results for grapevine were obtained by diluting 7.5 μg in 130 μL of EB. Avoid leaving the nucleic acid in the fragmentation tube for long periods. Transfer the eluate into the Covaris tube just before the sonication treatment, and subsequently move it immediately to a low-bind Eppendorf 1.5 mL tube. All equipment must be ice cooled. Also, avoid bubbles when transferring from one tube to another, and avoid shaking and aspirating slowly.
8. You can leave a bit of liquid to avoid carryover, because the remained liquid will not interfere.

9. Each time the tubes are rotated, they should be fully removed, twisted away from the magnet, and then placed back to ensure that all the beads pass through the EtOH solution. This step is essential for proper bead resuspension and efficient washing.
10. This process usually takes 10 min, but, depending on the number of beads, check every 5 min and remove any cracked samples immediately (to prevent excessive cracking).
11. At first, leave the tube on the magnetic rack and release the EB buffer from the tip directly on the beads, and then hold the tube with the hand and pipette up and down.
12. To reduce bead carryover when you are transferring the elution into a new tube, place the empty tube on the magnetic rack and use a p10 pipette with an elongated tip. Holding the tip against the magnet, slowly dispense the elution from the pipette into the bottom of the tube. The beads should visibly gather in the pipette tip near the magnet. Leaving 0.5-1 μL of liquid in the tip will prevent transfer of the beads.
13. Mix all the reagents together and then aliquot to the previously prepared samples. Since 2025, the End-It kit (Lucigen) is no longer provided—an alternative kit (Fast DNA End Repair Kit, ThermoFisher, Cat.no. K0771) could be considered: Add 5 μL of 10× End repair Reaction Mix, 2.5 μL of End Repair Enzyme Mix, and nuclease-free water to DNA samples, reaching 50 μL reaction volume. Incubate at 20 °C for 18 min (do not exceed 20 min).
14. Improved purification can be achieved by adding 10 μL of 3 M NaOAc, pH 5.5, to every 250 μL of PB buffer.
15. Mix all the reagents together and then aliquot to the previously prepared samples.
16. Better add the T4 DNA Ligase to each sample individually to avoid adaptor dimers.
17. This step is not strictly required but is included in ampDAP-Seq to remove DNA methylation marks. Additionally, it increases the DNA library yield and can serve as a control for fragmentation, either as an alternative or complementary to fragment analysis. It is important to ensure that the adapters used in this step are the same as those employed in the PCR enrichment step following the DAP reaction.
18. It is advisable to process eight samples (or a multiple thereof) at a time, primarily to prevent repeated freeze-thaw cycles of the TNT Coupled Reticulocyte Lysate, which can affect its performance. We recommend premixing all reagents—except for the pIX-Halo-ORF—directly in the TNT Coupled Reticulocyte Lysate tube before aliquoting the final mixture. Care should

be taken to avoid bubble formation during pipetting to ensure optimal reaction conditions.

19. We aliquot 39 μL, instead of 40 μL, to account for the residual bubbles that inevitably form during pipetting.
20. After this step, avoid placing the samples on ice, as this may cause precipitate formation. Additionally, it is highly recommended to proceed immediately with the DAP steps to prevent potential yield issues.
21. It is recommended to introduce a single large bubble into each well to ensure proper and uniform movement of the beads, preventing their stratification on either the cap or the bottom of the tube. This stratification must be strictly avoided, as it would result in protein binding only to the beads located at the interface between the stratified layers and the solution, leading to significant loss of the newly expressed transcription factor. Additionally, to maintain consistent and complete bead migration throughout the well, the rotation movement should be circular and slow.
22. In *Arabidopsis thaliana* (with a genome size of approximately 125 MB), different amounts of gDNA library, ranging from 40 to 100 ng, have been used, with the best results observed using 50–100 ng of library per transcription factor [6]. Based on these findings, we recommend adjusting the amount of gDNA library depending on the genome size of the species of interest. For instance, given that the grapevine genome is approximately 500 MB, an optimal range would be 200–500 ng per library. Furthermore, considering an average loss of 50% of gDNA during the library preparation process, along with the potential heterozygosity in some crop species, the required starting material should be approximately 400–1000 ng of genomic DNA per transcription factor. For ampDAP, the required amount of gDNA could be significantly lower.
23. This step will release DNA and denature proteins.
24. It is possible to store the samples at −20 °C, but it is highly recommended to follow the next step as soon as you finish this part.
25. This step attaches the TruSeq Illumina indexes for multiplexing sequencing and enriches the quantity of the final DNA fragments.
26. A second 1× AMPure beads cleanup can be performed, if necessary, to remove eventual adapter dimers or other unspecific dimers seen on the fragment analysis of the DAP-seq sample library. The Bioanalyzer results show an example for grapevine libraries, showing a well-defined peak shape (Fig. 1a), a "pure" DAP-seq sample library (Fig. 1b), and a

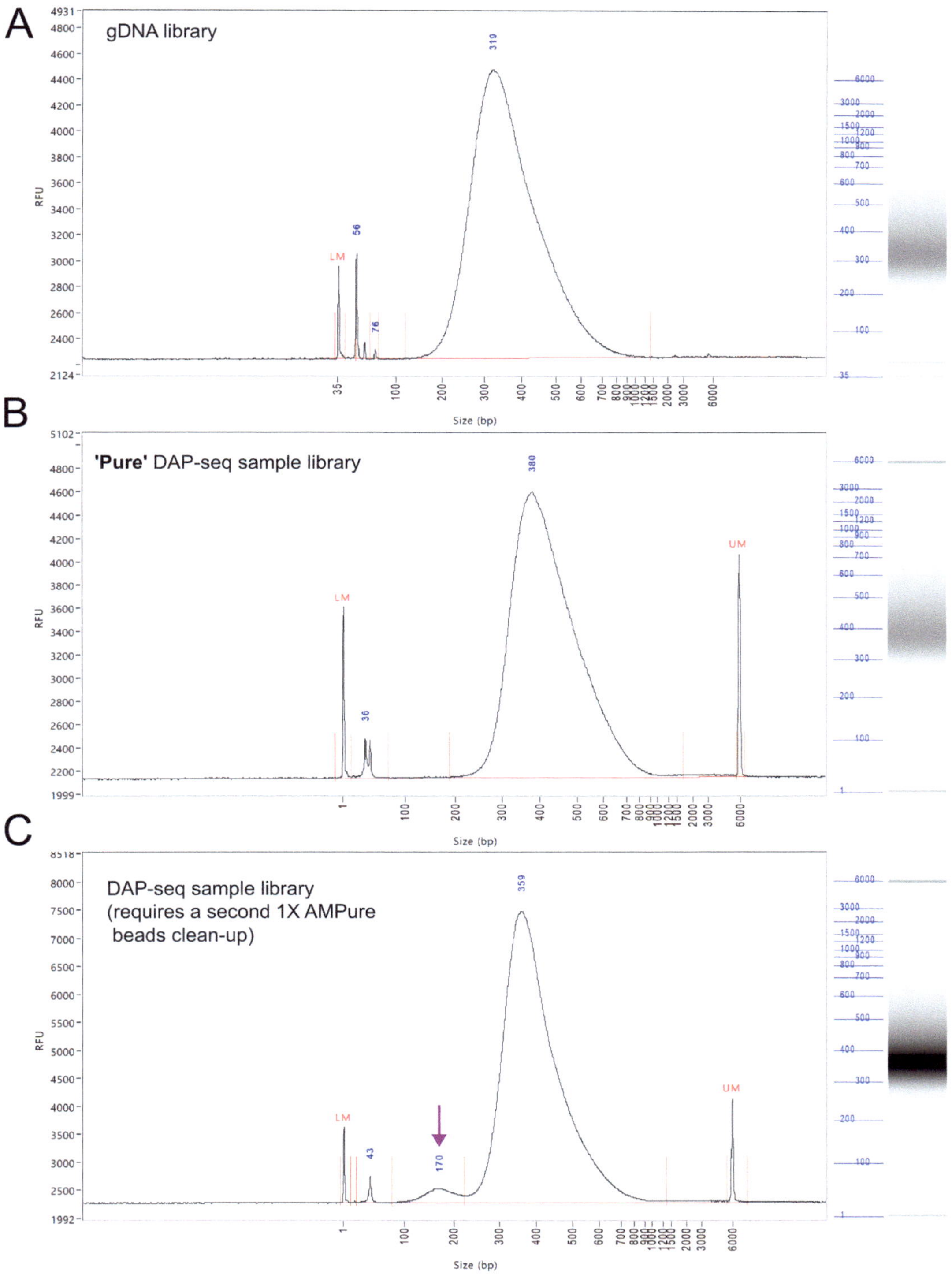

Fig. 1 Bioanalyzer results of an example for grapevine libraries, showing a well-defined peak shape (**a**), a "pure" DAP-seq sample library (**b**), and a DAP-seq sample library containing dimers (**c**)

DAP-seq sample library containing dimers (Fig. 1c). Dimers around 30–40 bp can be cleaned in the first AMPure XP beads cleaning, but those larger than 100 bp can affect the results and must be eliminated through a second cleanup. There are two approaches for this using the 1× AMPure beads—removing dimers from each sample individually or first preparing the sequencing pool library and then removing adapter dimers from the pooled library. For the preparation of sequencing pool library, it will depend on the sequencing platform.

27. The use of at least two replicates is crucial, as it enhances the ability to detect peaks in regions where a single-replicate analysis may fail. While the biological signal is typically consistent, TF libraries may contain technical noise, which is random in nature. By combining multiple TF libraries into a larger dataset for comparison against the joined input, the impact of this technical noise can be minimized. However, it is important to note that a true biological replicate does not exist in DAP-seq. This technique uses in vitro expressed transcription factor proteins to probe DNA libraries made from naked genomic DNA, making it challenging to define biological replicates in the traditional sense. The replicates in DAP-seq experiments involve genomic DNA fragments tested against the TFs in separate binding reactions, making them individual, independent gDNA-TF interaction samples. While they are not biological replicates, they can be considered technical replicates from the perspective of biological experiments. Replicate correlation in *Arabidopsis* DAP-seq has been shown to range from 0.70 to 0.99, with most correlations exceeding 0.9, whereas typical ChIP-seq replicate correlations range from 0.6 to 0.7. The reproducibility of DAP-seq experiments has also been validated in grapevine, showing high reproducibility (Spearman >0.8) [17].

28. The effect of sequencing depth on the detection of binding events (i.e., peaks) generally follows an asymptotic curve, where increasing depth initially increases the number of peaks, but this increase slows down and eventually plateaus. Unless the number of detected peaks is extremely low, we do not recommend sequencing beyond 50 M reads. Additionally, the appropriate sequencing depth should be chosen based on the size of the genome being studied.

29. For highly heterozygous and haplotype-resolved genomes, it is recommended to use each haplotype independently to prevent the loss of aligned reads due to multimapping. This approach ensures more accurate read mapping and avoids complications arising from the presence of multiple similar sequences.

30. Multireplicate analysis is always preferred over single-replicate analysis, as it increases the reliability and robustness of the results, helping to account for variability and improve the detection of true biological signals. As explained in the GEM peak caller tutorial (https://groups.csail.mit.edu/cgs/gem/), in a multireplicate analysis, GEM Peak caller combines the replicates as one large dataset, which means that two TF and input replicates are analyzed together. GEM peak caller uses the multireplicate mode by taking into account reads from both replicates and comparing them to the two input replicates to perform the peak calling. As described in its manual (https://groups.csail.mit.edu/cgs/gem/), GEM peak caller combines the replicates (pooling the reads) as one large dataset for the analysis. This approach is recommended by the ENCODE standard processing guidelines for ChIP-seq replicates as part of the Irreproducible Discovery Rate (IDR) pipeline.
31. The --*q* parameter is the *q-value* significance threshold for the detected peaks, specified as -log10(*q-value*). The default value is 2, which actually corresponds to a *q-value* threshold of 0.01. If a low number of peaks is detected, the --*q* parameter can be decreased, reducing the *q-value* threshold. For instance, for a 0.05 *q*-value threshold, a --*q* parameter of 1.3 should be chosen.
32. De novo motif discovery should be performed using 200 bp sequences centered at GEM-identified binding events for the 600 most enriched (e.g., best scoring) peaks.

Acknowledgments

This publication is based upon work from COST Action CA17111 INTEGRAPE and COST Innovators Grant IG17111 GRAPEDIA, supported by COST (European Cooperation in Science and Technology). This work was also supported by the NIH award R35GM138143 to S.C.H. and NSF Plant Genome Research Project grant IOS-1916804 to S.C.H. and by Grants PGC2018-099449-A-I00 and PID2021-128865NB-I00 from the Ministerio de Ciencia, Innovación y Universidades (MCIU, Spain), Agencia Estatal de Investigación (AEI, Spain), and Fondo Europeo de Desarrollo Regional (FEDER, European Union) awarded to J.T.M. S.Z. and A.V. are supported by PRIN2022 Project Number P2022AHJ99: "Vine-SCROLL: an innovative biotechnological approach for the in Situ Crispr-affinity puRification Of reguLatory eLements in grapevine." C.Z. was supported by China Scholarship

Council (CSC) no. 201906300087. This material was also in part based on work supported by the Center for Bioenergy Innovation (CBI), U.S. Department of Energy, Office of Science, Biological and Environmental Research Program under Award Number ERKP886 and the Oak Ridge National Laboratory Director's R&D (DRD) Program (M.L). Oak Ridge National Laboratory is managed by UT-Battelle, LLC for the Office of Science of the U.S. Department of Energy under Contract Number DE-AC05-00OR22725.

Conflict of Interest The authors have no conflicts of interest to declare that are relevant to the content of this chapter.

References

1. Lehti-Shiu MD, Panchy N, Wang P et al (2017) Diversity, expansion, and evolutionary novelty of plant DNA-binding transcription factor families. Biochim Biophys Acta Gene Regul Mech 1860:3–20. https://doi.org/10.1016/J.BBAGRM.2016.08.005
2. Jin J, Tian F, Yang D-C et al (2017) PlantTFDB 4.0: toward a central hub for transcription factors and regulatory interactions in plants. Nucleic Acids Res 45:D1040–D1045. https://doi.org/10.1093/nar/gkw982
3. Chow C-N, Lee T-Y, Hung Y-C et al (2019) PlantPAN3.0: a new and updated resource for reconstructing transcriptional regulatory networks from ChIP-seq experiments in plants. Nucleic Acids Res 47:D1155–D1163. https://doi.org/10.1093/nar/gky1081
4. Park PJ (2009) ChIP-seq: advantages and challenges of a maturing technology. Nat Rev Genet 10:669–680. https://doi.org/10.1038/nrg2641
5. O'Malley RC, Huang SC, Song L et al (2016) Cistrome and epicistrome features shape the regulatory DNA landscape. Cell 165:1280–1292. https://doi.org/10.1016/j.cell.2016.08.063
6. Bartlett A, O'Malley RC, Huang SC et al (2017) Mapping genome-wide transcription-factor binding sites using DAP-seq. Nat Protoc 12:1659–1672. https://doi.org/10.1038/nprot.2017.055
7. Galli M, Khakhar A, Lu Z et al (2018) The DNA binding landscape of the maize AUXIN RESPONSE FACTOR family. Nat Commun 9:1–14. https://doi.org/10.1038/s41467-018-06977-6
8. Song S, Huang B, Pan Z et al (2022) The SlTPL3-SlWUS module regulates multi-locule formation in tomato by modulating auxin and gibberellin levels in the shoot apical meristem. J Integr Plant Biol 64:2150–2167. https://doi.org/10.1111/jipb.13347
9. López-Vidriero I, Godoy M, Grau J et al (2021) DNA features beyond the transcription factor binding site specify target recognition by plant MYC2-related bHLH proteins. Plant Commun 2:100232. https://doi.org/10.1016/j.xplc.2021.100232
10. Santo Domingo M, Orduña L, Navarro D et al (2024) The ethylene-responsive transcription factor ERF024 is a novel regulator of climacteric fruit ripening in melon. Plant J 119:1844–1858. https://doi.org/10.1111/tpj.16889
11. Si J, Fan Z-Q, Wu C-J et al (2024) MaHsf24, a novel negative modulator, regulates cold tolerance in banana fruits by repressing the expression of HSPs and antioxidant enzyme genes. Plant Biotechnol J 22:2873–2886. https://doi.org/10.1111/pbi.14410
12. Zhu LJ, Gazin C, Lawson ND et al (2010) ChIPpeakAnno: a bioconductor package to annotate ChIP-seq and ChIP-chip data. BMC Bioinformatics 11:237. https://doi.org/10.1186/1471-2105-11-237
13. Langmead B, Salzberg SL (2012) Fast gapped-read alignment with bowtie 2. Nat Methods 9:357–359. https://doi.org/10.1038/nmeth.1923
14. Thomas MR, Matsumoto S, Cain P, Scott NS (1993) Repetitive DNA of grapevine: classes present and sequences suitable for cultivar

identification. Theor Appl Genet 86:173–180. https://doi.org/10.1007/BF00222076

15. D'Incà E, Foresti C, Orduña L et al (2023) The transcription factor VviNAC60 regulates senescence- and ripening-related processes in grapevine. Plant Physiol 192:1928–1946. https://doi.org/10.1093/plphys/kiad050
16. Fontanet-Manzaneque JB, Laibach N, Herrero-García I et al (2024) Untargeted mutagenesis of brassinosteroid receptor SbBRI1 confers drought tolerance by altering phenylpropanoid metabolism in Sorghum bicolor. Plant Biotechnol J 22:3406–3423. https://doi.org/10.1111/pbi.14461
17. Orduña L, Li M, Navarro-Payá D et al (2022) Direct regulation of shikimate, early phenylpropanoid and stilbenoid pathways by subgroup 2 R2R3-MYBs in grapevine. Plant J. https://doi.org/10.1111/tpj.15686

Chapter 2

Chromatin Immunoprecipitation in Crops

Carmen Martín-Pizarro

Abstract

Chromatin immunoprecipitation (ChIP) is a well-established technique that allows for the identification of genomic regions where proteins such as transcription factors, histones, or other chromatin-associated proteins bind. Moreover, deep sequencing of immunoprecipitated DNA (ChIP-seq) provides unbiased genome-wide insights into these binding sites. This chapter describes a detailed ChIP protocol tailored for transcription factors in recalcitrant plant leaves, using the commercial strawberry, the octoploid species *Fragaria* × *ananassa*, as an example. This protocol ensures obtaining DNA of both high quality and quantity for subsequent library preparation for Illumina sequencing. Finally, this chapter also details how to conduct quantitative PCR to validate the enrichment of the TF binding sites.

Key words Chromatin immunoprecipitation, ChIP, Antibodies, Transcription factors, Crops, Strawberry

1 Introduction

Chromatin comprises proteins essential for controlling nuclear processes such as gene transcription. Understanding the role of these proteins in chromatin regulation requires techniques that allow the identification of their DNA-binding sites. Among the various methods that have been developed, chromatin immunoprecipitation (ChIP) enables the in vivo identification of binding sites for DNA-binding proteins, generally histones or transcription factors (TFs), in specific DNA regions. The first ChIP protocol was developed in the mid-1980s to study the association of RNA polymerase II with DNA in *Escherichia coli* and *Drosophila* by Gilmour and Lis [1, 2]. Later in 1988, Solomon and colleagues used for the first time formaldehyde for protein-DNA cross-linking in *Drosophila* [3], becoming widely adopted since then for most organisms.

The ChIP assay includes the following steps: (1) It typically involves a first step of cross-linking with formaldehyde to fix DNA to proteins (X-ChIP). Alternatively, there are approaches without the cross-linking step (native chromatin immunoprecipitation,

José M. Franco-Zorrilla and Carmen Martín-Pizarro (eds.), *Plant Transcription Factors: Methods and Protocols*, Methods in Molecular Biology, vol. 2985, https://doi.org/10.1007/978-1-0716-4972-5_2,

N-ChIP): (2) isolation and (3) fragmentation (shearing) of chromatin, (4) immunoprecipitation of chromatin fragments using an antibody against the protein of interest (POI), (5) reverse cross-linking to recover DNA bound to the POI, and (6) identification of the binding sites. These binding sites can be determined in a targeted manner by qPCR. However, several years after the establishment of the ChIP method, genome-wide methods were developed to identify the protein binding sites. Firstly, the ChIP-chip assay was described, allowing the identification of binding sites using whole-genome microarrays [4]. Then the ChIP-seq arose, in which the binding sites are identified and sequenced by high-throughput sequencing [5]. This latest approach has become a commonly employed tool, offering a comprehensive view of the intricate regulatory mechanisms involving TFs and histones.

The ChIP method has been widely used in *Arabidopsis thaliana* and other species such as *Solanum lycopersicum* [6], *Oryza sativa* [7], and *Zea Mays* [8], among others. However, its application to recalcitrant and polyploid crops, such as strawberry, remains very limited. Nevertheless, recent advances in recalcitrant crops have been reported. In 2020, Hung and collaborators [9] conducted the first work on the diploid strawberry using the native (N-ChIP) methodology, which, as previously mentioned, is based on the immunoprecipitation of the native chromatin without cross-linking, followed by a micrococcal nuclease digestion (MNase) to fragment the chromatin. More recently, another protocol based on N-ChIP and X-ChIP, where the authors study the methylation of H3 histone in peach, has been published [10]. However, to date, it is difficult to find an X-ChIP protocol for crops suitable for subsequent ChIP-qPCR or ChIP-seq experiments studying TFs. As previously mentioned, the X-ChIP involves fixation with formaldehyde to cross-link the proteins to DNA, keeping them in in vivo conditions. Nevertheless, it presents several critical points when the goal is to identify direct targets for a particular TF, instead of histones. Firstly, it requires the generation of stable transgenic lines overexpressing the TF of interest fused with tag-proteins like GFP, HA, or FLAG, among others. In the case of recalcitrant crops, this process can vary from months to years, starting with the stable transformation until the transgenic lines are established in the greenhouse. Another critical point is the selection of appropriate antibodies against the POI, which is crucial for X-ChIP and must provide several advantages in specificity and sensitivity. Although the use of tagged-protein samples has the advantage of employing commercial antibodies with proven efficiency for this approach, the utilization of efficient antibodies specific to the native protein offers the advantage of eluding potential artifacts that may arise from protein tagging, potentially altering its functionality. This chapter provides a detailed explanation of an X-ChIP protocol that has been successfully applied to elucidate the comprehensive DNA-binding preferences of TFs in leaves of the octoploid strawberry *Fragaria* × *ananassa*. Additionally, it includes the verification of the TF's targets by qPCR. Importantly, this protocol yields the requisite amount of DNA for high-throughput sequencing of the targets (ChIP-seq).

2 Materials

All the stock solutions are prepared in ultrapure water and autoclaved or filtered with syringe filters with 22 μm pore size, before starting the experiment.

2.1 Tissue Harvest and Cross-Linking

1. Liquid nitrogen.
2. 0.5–1 g of plant tissue (*see* **Note 1**).
3. Mortars and pestles prechilled with liquid nitrogen and 50 mL Falcon tubes on ice.
4. Extraction buffer 1 freshly prepared and cold. Stock solutions: 1 M sucrose (filtered), 1 M HEPES, pH 7.5 (filtered), 0.2 M phenylmethylsulfonyl fluoride (PMSF, filtered), 1 M $MgCl_2$ (autoclaved), 5% polyvinylpyrrolidone (PVP-40, autoclaved), 14.3 M ß-mercaptoethanol, one tablet of complete protease inhibitor cocktail (Roche) in 50 mL.
5. Fixation buffer (freshly prepared and kept on ice): Extraction buffer 1 including 16% formaldehyde (*see* **Notes 2** and **3**).
6. 2 M glycine stock solution.

2.2 Chromatin Extraction and Shearing

1. 50 mL Falcon tubes, crystal funnels (autoclaved), pipette controller, 10 mL sterile pipettes.
2. Miracloth (pore size, 22–25 μm, Calbiochem).
3. Refrigerated centrifuge for 50 mL tubes.
4. Extraction buffer 2 freshly prepared. Stock solution: 1 M sucrose (filtered), 1 M HEPES, pH 7.5 (filtered), 0.2 M PMSF (filtered), 1 M $MgCl_2$ (autoclaved), 5% PVP-40 (autoclaved), 20% Triton X-100 (filtered), 14.3 M ß-mercaptoethanol, one tablet of complete protease inhibitor cocktail in 50 mL of Extraction buffer 2.
5. 15 mL Falcon tubes and refrigerated centrifuge for 15 mL tubes.
6. Extraction buffer 3, cold. Stock solution: 2 M sucrose (filtered), 1 M HEPES, pH 7.5 (filtered), 0.2 M PMSF (filtered), 1 M $MgCl_2$ (autoclaved), 5% PVP-40 (autoclaved), 20% Triton X-100 (filtered), 14.3 M ß-mercaptoethanol, one tablet of complete protease inhibitor cocktail in 50 mL of Extraction buffer 3.

7. 1.5 mL Eppendorf tubes and refrigerated centrifuge for 1.5 mL tubes.
8. Nuclei lysis buffer, room temperature. Stock solution: 1 M Tris-HCl, pH 8 (autoclaved), 0.5 M EDTA (autoclaved), 10% SDS (filtered), one tablet of complete protease inhibitor cocktail in 50 mL of nuclei lysis buffer.
9. BioruptorTM UCD-200 sonicator (Diagenode).

2.3 Chromatin Immunoprecipitation

1. Cold room and rotating wheel.
2. ChIP dilution buffer, cold. Stock solution: 1 M Tris-HCl. pH 8 (autoclaved), 5 M NaCl (autoclaved), 0.5 M EDTA (autoclaved), 20% Triton X-100 (filtered).
3. Protein A/G PLUS-Agarose beads (Santa Cruz Biotechnology) (*see* **Note 4**).
4. DNA LoBind 2 mL tubes.
5. Refrigerated microcentrifuge.
6. Low-salt wash buffer, cold. Stock solution: 1 M Tris-HCl, pH 8 (autoclaved), 0.5 M EDTA (autoclaved), 5 M NaCl (autoclaved), 10% SDS (filtered), 20% Triton X-100 (filtered).
7. High-salt wash buffer, cold. Stock solution: 1 M Tris-HCl, pH 8 (autoclaved), 0.5 M EDTA (autoclaved), 5 M NaCl (autoclaved), 10% SDS (filtered), 20% Triton X-100 (filtered).
8. LiCl wash buffer, cold. Stock solution: 1 M Tris-HCl, pH 8 (autoclaved), 0.5 M EDTA (autoclaved), 4 M LiCl (autoclaved), 10% sodium deoxycholate (filtered), 10% NP-40 substitute.
9. TE buffer, cold. Stock solution: 1 M Tris-HCl, pH 8 (autoclaved), 0.5 M EDTA (autoclaved).
10. ChIP elution buffer (filtrated), room temperature. Stock solution: 20% SDS, 0.168 g $NaHCO_3$ in 10 mL final volume.

2.4 Reverse Cross-Linking and DNA Purification

1. Thermo-mixer and microcentrifuge.
2. 5 M NaCl
3. RNase A/T1 Mix (Thermo Fisher).
4. Proteinase K, recombinant PCR Grade (Roche).
5. MinElute Reaction Cleanup Kit (Qiagen).

2.5 Validation of the Sample Enrichment

1. Primers for the genomic regions of interest and a negative control.
2. SsoFast EvaGreen Supermix for quantitave PCR (Bio-Rad).
3. Optical PCR plates for real-time PCR.

4. Real-time PCR machine.
5. Qubit fluorometer (Thermo Fisher Scientific™).

3 Methods

All the steps should be carried out on a clean bench, wearing gloves, with most of them in cold conditions, using always sterilized material and solutions.

3.1 Tissue Harvest and Cross-Link

1. Collect the tissue directly from the greenhouse in liquid nitrogen. Samples can be stored at −80 °C until use.
2. Grind the tissue with a cold mortar and pestle and collect the ground tissue in a prechilled 50 mL Falcon tube.
3. Weigh 0.5–1 g of powder in a new prechilled 50 mL Falcon tube. Add 25 mL of cold Fixation buffer, freshly prepared: 0.4 M sucrose (filtered), 10 mM HEPES pH 7.5 (filtered), 0.1 mM PMSF (filtered), 10 mM $MgCl_2$ (autoclaved), 1% PVP-40 (autoclaved), 5 mM ß-mercaptoethanol, one tablet of complete protease inhibitor cocktail, and 1% formaldehyde per tube—invert the tube several times to mix the sample with the buffer (*see* **Note 5**).
4. Cross-link the samples for 10 min at room temperature.
5. Stop the fixation by adding fresh glycine to a 0.15 M final concentration. Invert the tubes several times, mix well, and then incubate at room temperature for 5–10 min.

3.2 Chromatin Extraction and Shearing

1. Put an autoclaved funnel into the 50 mL Falcon tube on ice and a piece of Miracloth inside the funnel.
2. Wash the Miracloth filter several times using Extraction buffer 1 without formaldehyde to ensure the Miracloth is completely drenched.
3. Filter the samples through the wet Miracloth and collect them in a new 50 mL Falcon tube. Repeat the filtration two to three times to ensure clean nuclei extraction and debris elimination (*see* **Note 6**).
4. Centrifuge at 2300 × g for 20 min at 4 °C. Remove the supernatant by completely inverting the tube and keep the pellet.
5. Wash the nuclear pellet with 2 mL of chilled cold Extraction buffer 2 freshly prepared: 0.4 M sucrose (filtered), 10 mM HEPES, pH 7.5 (filtered), 0.1 mM PMSF (filtered), 10 mM $MgCl_2$ (autoclaved), 1% PVP-40 (autoclaved), 1% Triton X-100 (filtered), 5 mM ß-mercaptoethanol, one tablet of complete protease inhibitor cocktail—transfer the samples to a new

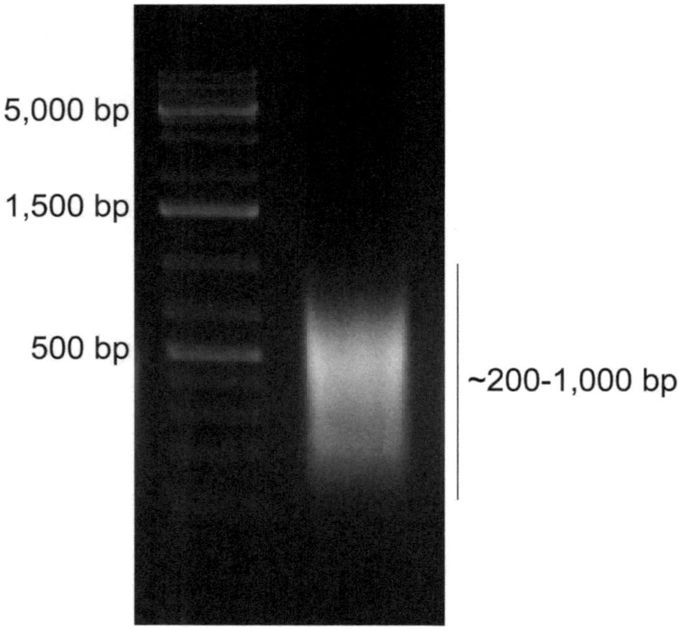

Fig. 1 DNA electrophoresis to test sonicated DNA using the shearing conditions depicted in this chapter for the Bioruptor® (Diagenode) sonicator. *Lane #1*: 1 kb GeneRuler Plus ladder (Thermo Fisher Scientific™). *Lane #2*: sonicated DNA

15 mL Falcon tube (*see* **Note 7**). Centrifuge at 13,500 × g for 10 min at 4 °C. Repeat this step four more times.

6. In a 1.5 mL tube, add 300 μL of cold Extraction buffer 3, freshly prepared: 1.7 M sucrose (filtered), 10 mM HEPES, pH 7.5 (filtered), 0.1 mM PMSF (filtered), 20 mM $MgCl_2$ (autoclaved), 1% PVP-40 (autoclaved), 0.15% Triton X-100 (filtered), 5 mM ß-mercaptoethanol, one tablet of complete protease inhibitor cocktail—use another 300 μL to resuspend the nuclear pellet in the 15 mL Falcon tube. Resuspend the pellet slowly, and then put it on top of the 300 μL of Extraction buffer 3 (*see* **Note 8**). Centrifuge at 18,000 × g for 1 h at 4 °C.
7. Discard the supernatant and gently resuspend the chromatin pellet in 250 μL of nuclei lysis buffer: 50 mM Tris-HCl, pH 8 (autoclaved), 10 mM EDTA (autoclaved), 1% SDS (filtered), one tablet of complete protease inhibitor cocktail by pipetting and/or vortexing (*see* **Note 9**).
8. Shearing the chromatin in a cold room. This step aims to shear the DNA into fragments from 200 bp to 1000 bp. The conditions depend on the sonicator, and it is recommended for the first time to check the size of the fragments in 2% agarose gel after the sonication (*see* **Note 10**) (Fig. 1).

9. Centrifuge the sonicated chromatin solution at 12,000 × g for 5 min at 4 °C to pellet the cellular debris.
10. Transfer the supernatant to a new 1.5 mL tube, and keep 20 μL of the supernatant at −80 °C. This will be used as the input sample (*see* **Note 11**).

3.3 Chromatin Immunoprecipitation

1. Dilute the chromatin sample ten times adding ChIP dilution buffer, cold: 20 mM Tris-HCl, pH 8 (autoclaved), 2 mM EDTA (autoclaved), 150 mM NaCl (autoclaved), 0.1% SDS (filtered), 1% Triton X-100 (filtered) to reduce the SDS concentration (*see* **Note 9**).
2. Put 50 μL of Protein A/G PLUS-Agarose beads into a 2 mL LoBind tube (*see* **Note 12**), add 1 mL of ChIP dilution buffer, and mix on a rotating wheel at 4 °C three times for 2 min to wash the beads. Centrifuge to pellet the beads at 3500 × g for 2 min at 4 °C. Discard the supernatant and keep the beads on ice.
3. Add the chromatin sample (approx. 2 mL) into the washed Protein A/G PLUS-Agarose beads, and incubate on a rotation wheel at 4 °C for 1 h. Centrifuge at 3500 × g for 2 min at 4 °C to pellet beads (*see* **Note 13**).
4. Transfer the supernatant containing the clean chromatin to the new 2 mL tube and add the antibody (depending on antibody concentration, usually around 1–2 μg in 1–5 μL diluted antibody solution).
5. Incubate the tubes with the chromatin and the antibody on a rotating wheel at 4 °C overnight.
6. Wash 50 μL of Protein A/G PLUS-Agarose beads in a 2 mL LoBind tube with 1 mL of ChIP dilution buffer, as in **step 2**. Centrifuge to pellet the beads at 3500 × g for 2 min at 4 °C and remove the supernatant. After that, add the chromatin with the antibody to this tube, and incubate it on a rotating wheel at 4 ° C for 1.5–2 h.
7. Centrifuge at 3500 × g for 2 min at 4 °C to pellet the beads (*see* **Note 14**).
8. Place the beads on a rotating wheel and washing them for 5 min at 4 °C, each time with 1 mL of one of the following washing buffers, in the following order—low-salt wash buffer: 20 mM Tris-HCl, pH 8 (autoclaved), 2 mM EDTA (autoclaved), 150 mM NaCl (autoclaved), 0.1% SDS (filtered), 1% Triton X-100 (filtered); high-salt wash buffer: High-salt wash buffer, cold: 20 mM Tris-HCl, pH 8 (autoclaved), 2 mM EDTA (autoclaved), 500 mM NaCl (autoclaved), 0.1% SDS (filtered), 1% Triton X-100 (filtered), LiCl wash buffer: 10 mM Tris-HCl,pH 8 (autoclaved), 1 mM EDTA (autoclaved),

250 mM LiCl (autoclaved), 1% sodium deoxycholate (filtrated), 1% NP-40 substitute, and finally with TE buffer: 10 mM Tris-HCl, pH 8 (autoclaved), 1 mM EDTA (autoclaved). After each wash, centrifuge at 3500 × g for 2 min at 4 °C to pellet beads, and discard the supernatant (*see* **Note 15**).

9. To elute the protein-DNA complexes from the beads, add 250 μL of freshly prepared ChIP elution buffer—1% SDS, 0.168 g $NaHCO_3$—vortex for 30 s, and incubate at 65 °C in agitation for 15 min.
10. Centrifuge the samples at 3500 × g for 2 min at room temperature, and transfer the supernatant to a new 1.5 mL tube. Repeat one more time this step, combining the elutions into the same tube for a final volume of 500 μL (*see* **Note 15**). This is the immunoprecipitated sample, or IP sample.

3.4 Reverse Cross-Linking and DNA Purification

1. From now on, include input sample (*see* **Notes 11** and **16**).
2. Add 20 μL 5 M NaCl to the IP (**step 10**, Subheading 3.3) and Input (**step 10**, Subheading 3.2) samples to begin the reverse cross-linking by incubating at 65 °C overnight.
3. Add 1 μL of RNase A/T1 (10 mg/mL) to each sample and incubate at 37 °C for 30 min.
4. Add 20 μL 1 M Tris-HCl, pH 6.5, 10 μL 0.5 M EDTA, and 2 μL Proteinase K (10 mg/mL) in the tubes, and incubate at 55 °C for 1 h.
5. Add another 1 μL of Proteinase K to the tubes and incubate at 37 °C for 1 h.
6. Purify IP and Input samples using MinElute Reaction Cleanup Kit following the manufacturer's protocol, with the following modification: Add 3× volume of ERC Buffer to each sample (*see* **Note 17**). Finally, elute the purification in 30 μL of Elution Buffer (EB). Repeat the elution step with flow through to maximize recovery.

3.5 Validation of the Sample Enrichment by qPCR

1. Dilute the DNA purified 1:6 in miliQ water.
2. Prepare the master mix for the positive and negative target loci as follows in a 10 μL final volume:

   0.5 μL 10 μM forward primer (final concentration: 500 nM) (*see* **Note 18**)

   0.5 μL 10 μM reverse primer (final concentration: 500 nM)

   4 μL diluted DNA sample

   5 μL EvaGreen Supermix.
3. Set up and run the following PCR program: initial denaturation for 10 min at 95 °C, followed by a total of 45 cycles of

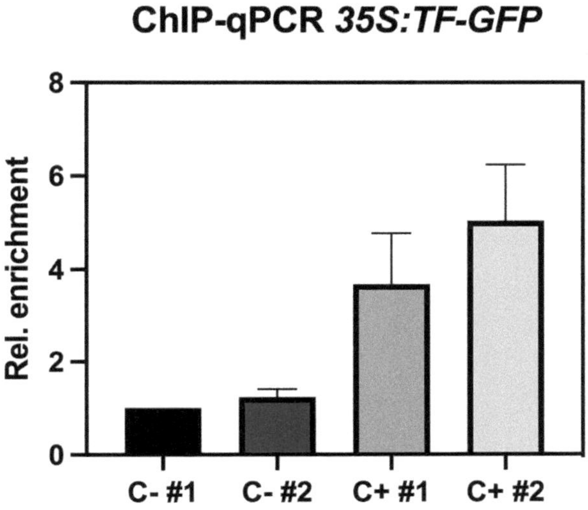

Fig. 2 ChIP-qPCR in a transgenic line overexpressing the TF of interest fused to GFP (*35S:TF-GFP*) showing around four- and fivefold enrichment for two positive target loci (C+ #1 and C+ #2). Data normalized using negative target loci (C- #1 and C- #2)

denaturation at 95 °C for 15 s and a single step of annealing and extension at 58 °C for 1 min.

4. qPCR analysis: Fold enrichment for the expected binding site is determined by comparing the cycle threshold (Ct) values obtained from immunoprecipitation (IP)-positive samples (tagged or wt samples) to those from IP-negative samples (untagged or mutant samples), prior normalization with their respective input DNA samples: ΔCt: Ct (IP-sample) − Ct (input-sample). Next, compare the positive and negative samples: ΔΔCt: ΔCt (positive sample) − ΔCt (negative sample). Then the enrichment is calculated: $2^{\Delta\Delta Ct}$. Finally, use the fold enrichment for the negative target locus to normalize the enrichment for the positive target loci (expected binding site) (*see* **Note 18**) (Fig. 2).
5. Once enrichment of positive target loci has been confirmed, DNA is ready to be quantified in a Qubit fluorometer (*see* **Note 19**).

4 Notes

1. Select the tissue and developmental stage where the transcription factor of interest is highly expressed. This protocol has been developed in leaves of strawberry plants. Fruit samples require optimization to obtain clean chromatin.

2. Prepare all the buffers freshly on the same experiment day using stock solutions. All the buffers must be prepared on ice (cold conditions).
3. Prepare more volume of Extraction buffer 1—we need to use it to wash and wet the Miracloth before passing the samples.
4. Select Protein A/G PLUS-Agarose beads depending on the species in which the primary antibody was generated.
5. Mix well the tissue with the Extraction buffer 1 to fix the sample properly.
6. Change the pipettes frequently to avoid introducing debris into the Miracloth and finally into the sample.
7. Resuspend the pellet gently using a pipette to avoid applying strong forces that could potentially damage the chromatin. Throughout the consecutive washes, the pellet typically lightens in color as a result of chloroplast washout.
8. In this step, the sucrose cushion produces a denser layer that, together with the high speed, reduces the contaminants of the sample and yields highly pure and intact nuclei.
9. It is important not to put more than 270 μL because the samples have to be later diluted ten times with ChIP dilution buffer, as indicated in Subheading 3.3 Chromatin Immunoprecipitation.
10. Sonication conditions. Bioruptor® (Diagenode): Three pulses for 20 min divided into 5–10-5 min; each pulse consists of 30 s with and 30 s without sonication. Keep the sonicator bath cold by adding ice to prevent heating of the samples.
11. The Input sample will be used again from Subheading 3.4.
12. Mix the tube of Protein A/G PLUS-Agarose beads before pipetting to homogenize beads throughout the solution and cut off the tip of the tip.
13. This step allows cleaning the chromatin of proteins that might nonspecifically bind to the Protein A/G PLUS-Agarose beads.
14. Keep the supernatant in a new 1.5 mL tube as a post-binding fraction at −20 °C to allow repeating the IP in case of failure.
15. Caution: Removing the supernatant to avoid carrying over any beads.
16. The IP and the Input sample are treated equally until the end of the protocol.
17. Be careful with the initial pH; sometimes it is necessary to add sodium acetate (NaOAc) buffer to the samples to adjust pH.
18. The primers to confirm the enrichment should be suitable for qPCR assays (80–120 bp long amplicons). The primers for the expected target gene/s of our TF of interest (positive target

loci) have to be designed to amplify the genomic region/s expected to be bound by this TF in a way that the primers flank the expected binding site. It is also important to design primers for loci where we do not expect any binding of our TF (negative target loci), being therefore our negative control to calculate the fold enrichment.

19. 3 ng of total DNA in IP samples is enough for library preparation for ChIP-seq analysis.

Acknowledgments

CMP and the research were supported by POSTDOC21_00893 contract and UMA20-FEDERJA-093 grant (Junta de Andalucía, Spain), respectively.

Conflict of Interest Statement The authors have no conflicts of interest to declare that are relevant to the content of this chapter.

References

1. Gilmour DS, Lis JT (1984) Detecting protein-DNA interactions in vivo: distribution of RNA polymerase on specific bacterial genes. Proc Natl Acad Sci USA 81:4275–4279. https://doi.org/10.1073/pnas.81.14.4275
2. Gilmour DS, Lis JT (1985) In vivo interactions of RNA polymerase II with genes of Drosophila melanogaster. Mol Cell Biol 5:2009–2018. https://doi.org/10.1128/mcb.5.8.2009-2018.1985
3. Solomon MJ, Larsen PL, Varshavsky A (1988) Mapping protein-DNA interactions in vivo with formaldehyde: evidence that histone H4 is retained on a highly transcribed gene. Cell 53:937–947. https://doi.org/10.1016/s0092-8674(88)90469-2
4. Thibaud-Nissen F, Wu H, Richmond T et al (2006) Development of Arabidopsis whole-genome microarrays and their application to the discovery of binding sites for the TGA2 transcription factor in salicylic acid-treated plants. Plant J 47:152–162. https://doi.org/10.1111/j.1365-313X.2006.02770.x
5. Kaufmann K, Muino JM, Osteras M et al (2010) Chromatin immunoprecipitation (ChIP) of plant transcription factors followed by sequencing (ChIP-SEQ) or hybridization to whole genome arrays (ChIP-CHIP). Nat Protoc 5:457–472. https://doi.org/10.1038/nprot.2009.244
6. Ricardi MM, González RM, Iusem ND (2010) Protocol: fine-tuning of a chromatin immunoprecipitation (ChIP) protocol in tomato. Plant Methods 6:11. https://doi.org/10.1186/1746-4811-6-11
7. Malone BM, Tan F, Bridges SM, Peng Z (2011) Comparison of four ChIP-Seq analytical algorithms using rice endosperm H3K27 trimethylation profiling data. PLoS One 6: e25260. https://doi.org/10.1371/journal.pone.0025260
8. He G, Chen B, Wang X et al (2013) Conservation and divergence of transcriptomic and epigenomic variation in maize hybrids. Genome Biol 14:R57. https://doi.org/10.1186/gb-2013-14-6-r57
9. Huang X, Pan Q, Lin Y et al (2020) A native chromatin immunoprecipitation (ChIP) protocol for studying histone modifications in strawberry fruits. Plant Methods 16:10. https://doi.org/10.1186/s13007-020-0556-z
10. Canton M, Farinati S, Forestan C et al (2022) An efficient chromatin immunoprecipitation (ChIP) protocol for studying histone modifications in peach reproductive tissues. Plant Methods 18:43. https://doi.org/10.1186/s13007-022-00876-0

Chapter 3

Cleavage Under Targets and Tagmentation (CUT&Tag) Protocol for Plants

Xiao-Yuan Tao and Jia-Wei Wang

Abstract

CUT&Tag (Cleavage Under Targets and Tagmentation) is a powerful method for chromatin profiling based on an enzyme-tethering strategy. Compared with chromatin immunoprecipitation (ChIP) assay, CUT&Tag requires smaller amount of cell/nuclei input and generates data with higher signal-to-noise ratio, which has been widely applied in profiling histone modifications and protein-DNA interactions. Here, we describe a detailed protocol for bulk-cell CUT&Tag in plants and briefly introduced other derivative CUT&Tag methods.

Key words CUT&Tag, Plant, B-CUT&Tag, Nano-CUT&Tag, Single-cell, Multimodal

1 Introduction

Cleavage Under Targets and Tagmentation (CUT&Tag) is an enzyme-tethering method that provides efficient, high-resolution sequencing libraries for profiling histone modifications or protein-DNA interactions in vivo [1]. CUT&Tag uses a hyperactive Tn5 transposase-Protein A/Protein G (pA/G-Tn5) fusion protein loaded with sequencing adapters. In the CUT&Tag reaction, a specific histone modification or chromatin protein is bound in situ by a specific antibody, which then tethers the protein A/G-Tn5 transposase fusion protein. In the presence of Mg^{2+}, the transposase is activated and performs the tagmentation of chromatin in situ, generating DNA fragments with "pasted" adaptors for the following next generation sequencing (NGS) libraries preparation (Fig. 1). CUT&Tag generates high-quality data using input of 100–100,000 cells/nuclei. Bulk-cell CUT&Tag protocols have been developed for animal cells and plants [2–4]. Recently, multiple derivative CUT&Tag methods have been developed based on biotinylated pA/G Tn5 transposase for TF-chromatin interaction profiling in plants (B-CUT&Tag) [5] or based on a set of secondary

José M. Franco-Zorrilla and Carmen Martín-Pizarro (eds.), *Plant Transcription Factors: Methods and Protocols*, Methods in Molecular Biology, vol. 2985, https://doi.org/10.1007/978-1-0716-4972-5_3,

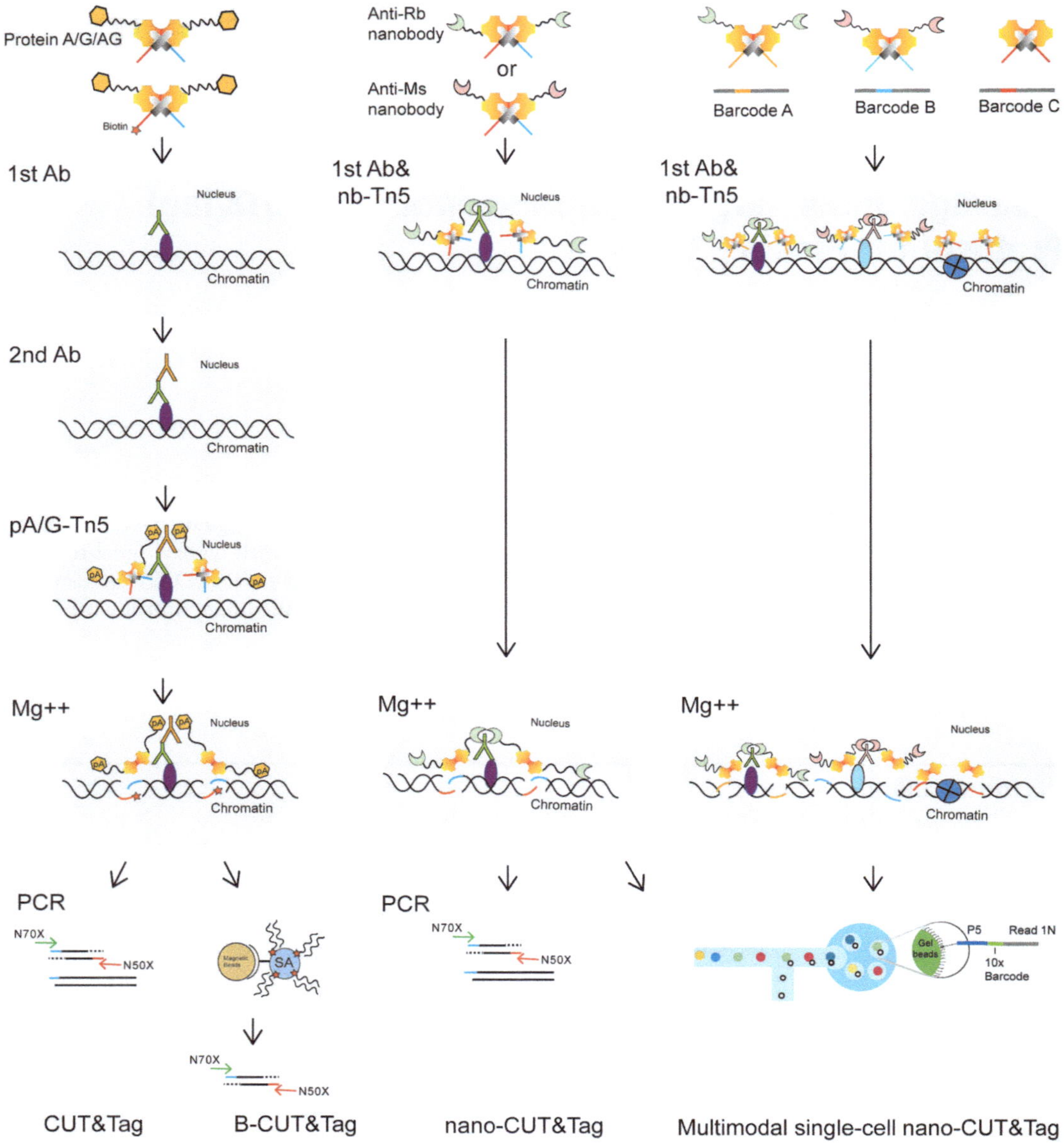

Fig. 1 Workflow of CUT&Tag and its derivative methods. Intact nuclei are used in plant CUT&Tag. For bulk-CUT&Tag, the primary antibody (green) binds to recognize the target chromatin protein (blue), and a second antibody (orange) is then added and enhances in situ tethering of pA/pG-Tn5 transposome. Transposome is activated by Mg++ and integrates adapters at chromatin protein binding sites. The resulting fragments with adapters at both ends are enriched by PCR and analyzed by NGS. Alternatively, the pA/pG-Tn5 can be assembled with biotinylated adaptors, enabling the generation of biotinylated target fragments for SA-biotin-based purification, followed by PCR and NGS (B-CUT&Tag). Nano-CUT&Tag uses the anti-mouse or anti-rabbit IgG secondary nanobody-fused Tn5. The primary antibody and nb-Tn5 are added together for in situ recognition and binding, resulting in fewer washing steps and a more concise procedure. Nano-Tn5 with different barcoding adaptors is used in multimodal nano-CUT&Tag and followed by single-cell profiling (multimodal & single-cell nano-CUT&Tag)

nanobody (e.g., anti-mouse IgG, anti-rabbit IgG) fused Tn5 transposase for profiling up to three targets in one reaction (multimodal CUT&Tag) [6] or accommodate single-cell (single-cell CUT&-Tag) and spatial systems (spatial CUT&Tag for heterogeneity profiling) [7, 8]. Here, we introduce a streamlined bulk-cell CUT&Tag protocol for efficient chromatin profiling in plants using intact nuclei.

2 Materials

2.1 Reagents and Equipment

1. Tn5 transposases (*see* **Note 1**)
   - Protein A/G-Tn5 transposase.
   - Anti-mouse nanobody-Tn5 transposase (Ms-Tn5, Addgene), for nano-CUT&Tag.
   - Anti-rabbit nanobody-Tn5 transposase (Rb-Tn5, Addgene), for nano-CUT&Tag.
2. Antibodies
   - Mouse anti-H3K4me3 antibody (ChIP Grade, 1 mg/mL).
   - Rabbit anti-H3K4me3 antibody (ChIP Grade, 1 mg/mL).
   - Mouse IgG, used as a control antibody.
   - Secondary antibody: Rabbit Anti-Mouse IgG; Guinea Pig anti-Rabbit IgG.
3. Oligos
   - ME-rev: 5′-CTGTCTCTTATACACATCT-3′, 5′--Phosphate, 3′-Aminolinker C7.
   - ME-A: 5′--TCGTCGGCAGCGTCAGATGTGTATAAGAGACAG-3′.
   - Biotinylated ME-A: 5′--TCGTCGGCAGCGTCAGATGTGTATAAGAGACAG-3′, 5′-Biotin TEG modification, for B-CUT&Tag.
   - ME-B: 5′--GTCTCGTGGGCTCGGAGATGTGTATAAGAGACAG-3′.
   - N501 to N508: AATGATACGGCGACCACCGAGATC-TACACNNNNNNNNTCGTCGGCAGCGTC, in which "NNNNNNNN" = TAGATCGC, CTCTCTAT, TATCCTCT, AGAGTAGA, GTAAGGAG, ACTGCATA, AAGGAGTA, and CTAAGCCT, respectively.
   - N701 to N712: CAAGCAGAAGACGGCATACGA-GATnnnnnnnnGTCTCGTGGGCTCGG, in which "nnnnnnnn" = TAAGGCGA, CGTACTAG, AGGCAGAA, TCCTGAGC, GGACTCCT, TAGGCATG, CTCTCTAC, CAGAGAGG, GCTACGCT, CGAGGCTG, AAGAGGCA, and GTAGAGGA, respectively.

4. Stock reagents
   - Annealing buffer: 10 mM Tris-HCl, pH 8.0, 50 mM NaCl, 1 mM EDTA. Stored at −20 °C.
   - 10× isolation buffer (10× IsoB): Add 5 mL 1 M Tris-HCl, pH 7.5, 5 mL 1 M KCl, 1 mL 0.5 M EDTA, 125 μL 2 M spermidine, and 200 μL β-mercaptoethanol, and fill with distilled water to a final volume of 50 mL. Store the buffer at 4 °C.
   - 10 × wash buffer (10 × WB): Add 1 mL 1 M Tris-HCl, pH 7.5, 15 mL 5 M NaCl, and 125 μL 2 M spermidine, and fill with distilled water to a final volume of 50 mL. Store the buffer at 4 °C.
   - 5 × antibody buffer (5 × AB): Add 1.25 mL 1 M Tris-HCl, pH 7.5, 50 μL 0.5 M EDTA, pH 8.5, 0.75 mL mL 5 M NaCl, and 6.25 μL 2 M spermidine, and 0.25 mL 100 mg/mL bovine serum albumin (BSA), and fill with distilled water to a final volume of 5 mL. Store the buffer at −20 °C.
   - Binding buffer: 20 mM Tris-HCl, pH 7.5, 10 mM KCl, 1 mM $MnCl_2$, 1 mM $CaCl_2$. Add 200 μL 1 M Tris-HCl, pH 7.5, 100 μL 1 M KCl, 10 μL 1 M $MnCl_2$, and 10 μL 1 M $CaCl_2$, and fill with distilled water to a final volume of 10 mL. Store the buffer at 4 °C.
   - 1 M $MgCl_2$; store at 4 °C.
   - 2 M glycine; store at 4 °C.
   - 20% Triton X-100; store at room temperature.
   - 20% Tween-20; store at room temperature.
   - 10% SDS; store at room temperature.
   - 5% digitonin (~50% TLC) in DMSO; store at −20 °C.
   - 37% formaldehyde.
   - Protease inhibitor cocktail for plant cell extracts (100×); store at −20 °C.
   - Phenol/chloroform/isoamyl alcohol (25:24:1, v:v:v); store at room temperature.
   - Chloroform.
   - CTAB DNA extraction buffer: 100 mM Tris-HCl, pH 8.0, 20 mM EDTA, pH 8.5,1.4 M NaCl, 2% (w/v) CTAB (cetyltrimethylammonium bromide), 1% PVP 40,000 (polyvinylpyrrolidone).
   - GlycoBlue Coprecipitant (Invitrogen, 15 mg/mL).
   - DAPI stock solution (1 mg/mL).
   - High-fidelity DNA polymerase enzyme for NGS.

- Concanavalin A-coated beads (Con-A beads), for nuclei binding, optional.
- SPRI DNA purification beads.
- Streptavidin magnetic beads, for B-CUT&Tag.
- 2 × Dialysis buffer: 100 mM Tris-HCl, pH 7.5, 200 mM NaCl, 0.2 mM EDTA, 0.2% Triton X-100, 20% glycerol, 2 mM DTT. For homemade nanobody-Tn5 transposase.

5. Equipment
   - NanoDrop spectrophotometer.
   - Centrifuge.
   - 500-mesh sieve.
   - Rocking shaker.
   - Magnetic stand for 1.5 mL tubes.
   - Bioanalyzer with appropriate kits.

2.2 Working Solutions

Prepare fresh working solutions before use.

1. Isolation buffer (IsoB): 10 mM Tris-HCl, pH 7.5, 10 mM KCl, 0.5 mM spermidine, 1 mM EDTA, β-mercaptoethanol. Add 5 mL 10× IsoB together with distilled water to a final volume of 50 mL.
2. Isolation buffer with 1% formaldehyde (IsoB (F+)): 10 mM Tris-HCl, pH 7.5, 10 mM KCl, 0.5 mM spermidine, 1 mM EDTA, β-mercaptoethanol, 1% formaldehyde. Add 0.7 mL 37% formaldehyde for each 25 mL IsoB.
3. Isolation buffer with Triton X-100 (IsoB (T+)): 10 mM Tris-HCl, pH 7.5, 10 mM KCl, 0.5 mM spermidine, 1 mM EDTA, β-mercaptoethanol, 0.5% Triton X-100. Add 250 μL 20% Triton X-100 (final 0.5%) and 10 μL protease inhibitor cocktail (final 0.1%) for each 10 mL IsoB.
4. Wash buffer with protease inhibitor cocktail (WB (C+)): 10 mM Tris-HCl, pH 7.5, 150 mM NaCl, 0.5 mM spermidine, 0.1% protease inhibitor cocktail. Add 2.5 mL 10 × WB and 25 μL protease inhibitor cocktail together with distilled water to a final volume of 25 mL.
5. Antibody buffer with protease inhibitor cocktail and digitonin (AB (C + D+)): 50 mM Tris-HCl, pH 8.0, 1 mM EDTA, 150 mM NaCl, 0.5 mM spermidine, 1 mg/mL BSA, 1% protease inhibitor cocktail, 0.05% w/v digitonin. Add 0.2 mL 5 × AB, 10 μL protease inhibitor cocktail (final 1%) and 10 μL 5% w/v digitonin together with distilled water to a final volume of 1 mL.

6. Wash buffer with protease inhibitor cocktail and digitonin (WB (C + D+)): 10 mM Tris-HCl, pH 7.5, 150 mM NaCl, 0.5 mM spermidine, 0.1% protease inhibitor cocktail, 0.05% digitonin. Add 5 mL 10 × WB, 50 μL protease inhibitor cocktail, and 500 μL 5% w/v digitonin together with distilled water to a final volume of 50 mL.
7. Transposase incubation buffer (TIB): 10 mM Tris-HCl, pH 8.0, 300 mM NaCl, 0.5 mM spermidine, protease inhibitor cocktail 0.1%, 0.05% w/v digitonin. Mix 1 mL WB (C + D+) with 50 μL 3 M NaCl.
8. Tagmentation buffer (TB): 10 mM Tris-HCl, pH 8.0, 300 mM NaCl, 0.5 mM spermidine, protease inhibitor cocktail 0.1%, 10 mM $MgCl_2$, 0.05% w/v digitonin. Mix 1.2 mL WB (C + D +), 60 μL 3 M NaCl, with 12 μL 1 M $MgCl_2$.
9. 2 × Western blot buffer: 10 mM Tris-HCl, pH 7.5, 1 mM EDTA, 2 M NaCl, 0.04% Tween-20. This buffer is used for B-CUT&Tag in subheading 3.6.

3 Methods

3.1 Tn5 Assembly

1. Add annealing buffer to oligos ME-rev, ME-A, and ME-B to make a 100 μM stock solution for each oligo.
2. Set up the following two reactions in two PCR tubes to anneal the adapters—Reaction 1 (ME-A/ME-rev): For a total volume of 20 μL, add 10 μL of 100 μM ME-rev and 10 μL of 100 μM ME-A. Reaction 2 (ME-B/ME-rev): For a total volume of 20 μL, add 10 μL of 100 μM ME-rev and 10 μL of 100 μM ME-B. Anneal the oligos using the program in the PCR machine (95 °C for 3 min, 75 °C for 15 min, 60 °C for 10 min, 50 °C for 10 min, 40 °C for 10 min, 25 °C for 30 min).
3. Mix the ME-A/ME-rev and ME-B/ME-rev at a 1:1 ratio, designated as the adapter mix.
4. Set up the following reaction to generate the transposase: For a total volume of 9.375 μL, add 5 μL of hyperactive pA/G-Tn5 transposase (500 ng/μL), 0.875 μL of adapter mix, and 3.5 μL of coupling buffer. Pipette gently 20 times and mix well; incubate at 30 °C for 1 h. The final concentration of pA/G-Tn5 transposase product is 4 pmol/μL. Store the transposase at − 20 °C.

3.2 Nuclei Isolation

1. For transcription factor profiling, transfer plant tissue (~ 0.5 g) to a 50 mL Falcon tube containing 25 mL of IsoB (F+), and vacuum infiltrate for 10 min at −25 inHg. For histone modification profiling, no cross-linking is needed; go to **step 4**.

2. Stop the cross-linking by the addition of 1.6 mL of 2 M glycine to a final concentration of 0.125 M, mix well, and continue vacuum infiltration for 5 min at −25 inHg.
3. Rinse cross-linked material with H_2O, and thoroughly remove any excess of water using paper towels.
4. Grind the fresh or cross-linked tissue (~1 g) in liquid nitrogen to a fine powder.
5. Transfer the powder to 25 mL chilled IsoB, mix gently, and incubate the tube on ice for 5 min with gentle shaking to disperse the material evenly.
6. Centrifuge for 5 min at 500×*g* at 4 °C.
7. Decant the supernatant, resuspend the pellet with 10 mL of chilled IsoB (T+) (*see* **Note 2**). Invert the tube gently to resuspend the pellet completely.
8. Filter the lysate using a 500-mesh sieve.
9. Centrifuge for 5 min at 300×*g* at 4 °C to collect the nuclei.
10. Remove as much of the supernatant as possible. Wash three times with 800 μL of WB (C+).
11. Resuspend 50 μL nuclei (estimated volume under 300×*g*) with 1 mL WB (C+). Divide into six 1.5 mL tubes with 150 μL nuclei each (~10,000–20,000 nuclei). Centrifuge for 3 min at 300×*g* at 4 °C. Remove as much of the remaining buffer as possible using a pipette tip.
12. (Optional) For QC of nuclei isolation, DAPI staining and imaging can be performed: Transfer 100 μL nuclei suspension to a new 1.5 mL tube, and add 900 μL WB (C+) and 2 μL DAPI stock solution (1 mg/mL) to a final working concentration of 2 μg/mL of DAPI. Incubate the tube in dark for 30 min at room temperature. After staining, centrifuge for 3 min at 300×*g* at 4 °C to collect the nuclei. Remove the supernatant as much as possible. Wash three times with 800 μL of WB (C+). Remove as much of the supernatant as possible, and resuspend the nuclei with 100 μL WB (C+). Image the nuclei under a fluorescence microscope using the DAPI channel (Fig. 2). Count the concentration of nuclei using manual counting chambers.

3.3 CUT&Tag Reaction

1. (Optional; *see* **Note 3**) Set up six tubes of replicates of the following washing: Wash 20 μL Con-A beads with 500 μL binding buffer twice to activate Con-A beads. Place the tube on a magnet stand and remove the liquid.
2. (Optional) Resuspend Con-A beads with 150 μL binding buffer. Add the activated Con-A beads to nuclei and incubate at 4 °C for 15 min.

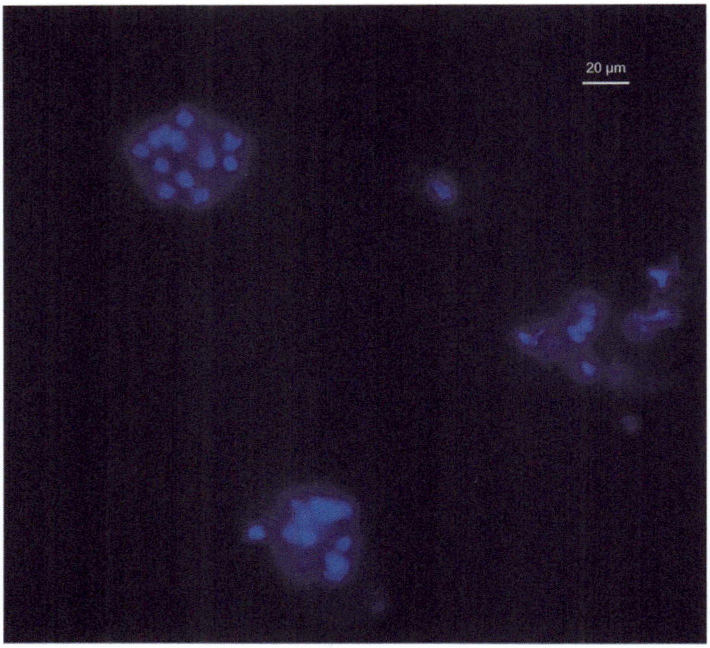

Fig. 2 Intact nuclei isolated from leaves of 4-week-old cotton plants. Nuclei were isolated as described in Subheading 3.2, **step 4–11** using IsoB(T+) with 0.5% Triton X-100. DAPI staining of nuclei was performed as described in Subheading 3.2, **step 12**. After staining, image was taken under a fluorescence microscope using the DAPI channel. Bar = 20 μm

3. (Optional) Discard the liquid and collect nuclei by a magnetic stand.
4. Resuspend the nuclei with 150 μL ice-cold AB (C + D+). Set up the following reactions in 1.5 mL tubes: two biological replicates of IgG control groups with 1 μL IgG (1 mg/mL) in each tube; two biological replicates of H3K4me3 assay groups with 1 μL anti-H3K4me3 antibody in each tube; and two biological replicates of H3K4me3 assay groups with 1 μL anti-H3K27me3 antibody in each tube. Mix the solutions gently, and place the tubes on a rocking shaker overnight at 4 °C and 12 rpm.
5. After antibody incubation, discard the liquid and collect nuclei by a magnetic stand or by centrifuge for 3 min at 300×*g* at 4 °C.
6. Resuspend the nuclei with 150 μL of WB (C + D+) with 1 μL secondary antibody, and leave the tubes on a rocking shaker for 1 h at room temperature and 12 rpm.
7. After secondary antibody incubation, discard the liquid and collect nuclei by a magnetic stand or by centrifugation for 3 min at 300×*g* at 4 °C.

8. Add 800 μL WB (C+D+) to each tube; mix gently. Leave the tubes on a rocking shaker for 5 min at room temperature and 12 rpm. Discard the liquid and collect nuclei by a magnetic stand or by centrifuge for 3 min at 300×*g* at 4 °C.
9. Repeat washing steps two more times.
10. After the third wash, discard the liquid and collect nuclei by a magnetic stand or by centrifuge for 3 min at 300×*g* at 4 °C. Remove the remaining buffer as much as possible.
11. Prepare transposase solution by adding 6 μL of transposase to 900 μL TIB, and mix gently.
12. Resuspend the nuclei with 150 μL of transposase solution. Place the tubes on a rocking shaker at 12 rpm for 2–3 h at room temperature.
13. After transposase incubation, discard the liquid and collect nuclei by a magnetic stand or by centrifuge for 3 min at 300×*g* at 4 °C.
14. Add 800 μL WB (C+D+) to each tube, mix gently, and leave the tubes on a rocking shaker at 12 rpm for 5 min at room temperature. Discard the liquid and collect nuclei by a magnetic stand or by centrifuge for 3 min at 300×*g* at 4 °C.
15. Repeat washing steps two more times.
16. After the third wash, discard the liquid and collect nuclei by a magnetic stand or by centrifuge for 3 min at 300×*g* at 4 °C. Remove the remaining buffer as much as possible.
17. Resuspend the nuclei with 300 μL of TB; incubate the tubes in a 37 °C water bath for 1 h for tagmentation.
18. Stop the tagmentation with 10 μL of 0.5 M EDTA (final concentration ~15 mM), 30 μL of 10% SDS (final concentration 1%), and 5 μL 20 mg/mL proteinase K. Incubate the tubes in a 50 °C water bath for 1 h.

3.4 DNA Extraction and Library Construction

1. Add 300 μL of CTAB DNA extraction buffer. Incubate the tubes in a 65 °C water bath for 30 min. Invert to mix every ~10 min.
2. Add 600 μL phenol/chloroform/isoamyl alcohol (25:24:1) to each tube. Mix thoroughly.
3. Pre-centrifuge the phase lock gel tubes for 2 min at 13,000×*g* at 4 °C. Transfer the above solution to phase lock gel tubes.
4. Centrifuge for 10 min at 13,000×*g* at 4 °C. Transfer the supernatant (~600 μL) to a new 1.5 mL tube.
5. Add 600 μL of chloroform to each tube. Mix thoroughly. Centrifuge for 10 min at 13,000×*g* at 4 °C.

6. Transfer the supernatant (~600 μL) to a new 1.5 mL tube. Add 1.2 mL of 100% ethanol and 2 μL of GlycoBlue Coprecipitant. Mix thoroughly and store at −20 °C for 1 h.
7. Centrifuge for 10 min at 13,000×*g* at 4 °C. Decant the supernatant. Wash the DNA pellet using 75% (v/v) ethanol, and centrifuge for 5 min at 13,000×*g* at 4 °C.
8. Decant the supernatant. Air-dry the DNA pellet and resuspend the DNA with 25 μL H_2O.
9. Set up the PCR reaction as follows: For a total of 50 μL reaction, add 24 μL DNA, 11 μL H_2O, 10 μL 5 × TAE buffer, 2 μL P5 (N50X; *see* **Note 4**) primer (10 μM), 2 μL P7 (N70X) primer (10 μM), 1 μL TruePrep Amplify Enzyme (TAE, Vazyme). PCR program: 72 °C for 3 min (*see* **Note 5**), 98 °C for 30 s and then 14–18 cycles of 98 °C for 30 s, 60 °C for 30 s, and 72 °C for 30 s, followed by 72 °C for 5 min.
10. Load 3 μL of the PCR products on 1.5% agarose gel for electrophoresis. Add one to two PCR cycles if needed.
11. Purify the PCR products using commercial SPRI DNA purification magnetic beads.
12. Quantification of library with Bioanalyzer and qPCR (Fig. 3). Effective concentration of the library should be >2 nM to ensure library quality.
13. Perform paired-end Illumina sequencing using an Illumina HiSeq 2500 or another massively parallel DNA sequencer, following the manufacturer's instructions. Obtain a 6- to 7-G raw base data/sample.

3.5 Data Analysis

1. Bioinformatic analysis can be performed as described [4]. Fastp v 0.20.1 is used to remove adapters and low-quality reads. Align paired-end reads using Hisat2 v 2.2.0 with the following parameters: --no-spliced-alignment --no-mixed --no-discordant --phred33 -I 10 -X 700. Unique aligned reads are extracted using Perl script: cat aligned.sam | perl -ne "print if /^@|NH:i:1\b/." Duplicated reads are removed using Picard v 2.22.8 (Picard Toolkit 2019, Broad Institute, GitHub Repository, http://broadinstitute.github.io/picard/) with this parameter: REMOVE_DUPLICATES = true. Peak calling uses macs2 v 2.1.3.3 with these parameters: macs2 callpeak -t input_file -p 1e-5 -f BAM -keep-dup all -n out_name. Scatterplots, correlation plots, and heatmaps are displayed using deepTools v 3.1.3. Annotation of peaks is performed using an R/Bioconductor package, ChIPseeker.

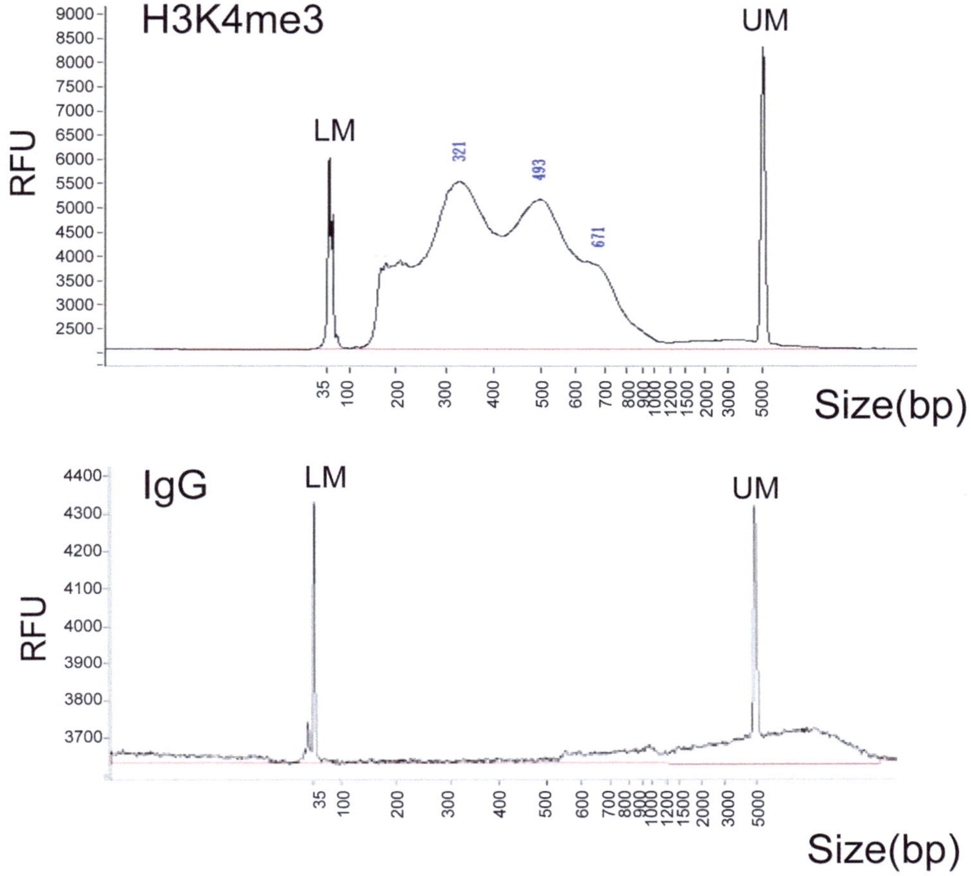

Fig. 3 Fragment analysis of CUT&Tag libraries using Bioanalyzer. Fragment distribution of representative H3K4me3 library generated using nuclei isolated from leaves of 4-week-old cotton plants. IgG library in CUT&Tag was used as a control. *LM* lower markers, *UM* upper markers, *RFU* relative fluorescence unit

3.6 B-CUT&Tag

B-CUT&Tag is performed similar to regular bulk-cell CUT&Tag as described in Subheadings 3.1, 3.3, and 3.4 with the following differences (Fig. 1) (*see* **Note 6**):

1. For B-CUT&Tag, in Subheading 3.1, prepare the adaptor mix with biotinylated ME-A instead for pA/pG-Tn5 assembly.
2. In **step 4** of Subheading 3.3, set up multiple technical replicates of CUT&Tag reaction, and scale up the reaction nuclei input/volume, e.g., 300 μL nuclei for each reaction with 2 μL primary antibody.
3. After **step 8** in Subheading 3.4, combine the resulting DNA from multiple technical replicates together for biotin/streptavidin affinity purification.

4. Wash twice a total of 20 μL of Streptavidin magnetic beads with 1 × Western blot buffer and once with 2 × Western blot buffer, and finally resuspend in 40–100 μL of 2 × Western blot buffer.
5. Add an equal volume of DNA sample to the beads, and incubate the mixture at room temperature for 30 min with rotation.
6. Extensively wash isolated DNA on beads three times with 1 × Western blot buffer, wash once with 10 mM Tris-HCl, pH 7.5 buffer, and finally resuspend in 25 μL of H_2O for subsequent PCR-based NGS library construction described in **step 9** of Subheading 3.4.

3.7 Nano-CUT&Tag

Compared with regular bulk-cell CUT&Tag as described in Subheadings 3.1, 3.3, and 3.4, nano-CUT&Tag has the following differences (Fig. 1) (*see* **Note 6**):

1. For the unbarcoded version (single histone modality) nano-CUT&Tag (Fig. 1), in Subheading 3.1, assemble anti-mouse or anti-rabbit nanobody-Tn5, same as that in bulk-cell CUT&-Tag (*see* **Note 6**).
2. In **step 4** of Subheading 3.3, set up the CUT&Tag reaction by adding primary antibody and nano-Tn5 together: two biological replicates of IgG control groups with 1 μL IgG (1 mg/mL) and 1 μL anti-mouse or anti-rabbit nanobody-Tn5 in each tube and two biological replicates of H3K4me3 assay groups with 1 μL anti-H3K4me3 antibody and 1 μL anti-mouse or anti-rabbit nanobody-Tn5 in each tube. Mix the solution gently, and incubate the tubes on a rocking shaker overnight at 4 °C and 12×*g*.
3. In nano-CUT&Tag, the steps of secondary antibody incubation, Tn5 incubation, and washing (**steps 5–12** in Subheading 3.3) are not needed.
4. For single-cell and multimodal nano-CUT&Tag, in Subheading 3.1, assemble anti-mouse and anti-rabbit nanobody-Tn5 with different barcoded ME-A/ME-rev adaptors as described (https://www.protocols.io/view/nano-cut-amp-tag-for-multimodal-profiling-of-the-c-8epv59o8dg1b/v1) [6], e.g., ME-A_bcd1/ME-rev adaptors assembled with anti-mouse nanobody-Tn5, ME-A_bcd2/ME-rev adaptors assembled with anti-rabbit nanobody-Tn5, and ME-A_bcd3/ME-rev adaptors assembled with Tn5 at 1:1 ratio in pmol.
5. For single-cell and multimodal nano-CUT&Tag, in **step 18** of Subheading 3.3, the tagmentation reaction of CUT&Tag is stopped by adding stop buffer with 50 mM EDTA but without SDS. Then it was accommodated to 10× genomics scATAC-seq approaches, following closely the Chromium Next GEM Single Cell ATAC Reagent Kits v1.1 manual (*see* **Note 7**).

4 Notes

1. For bulk-cell CUT&Tag and B-CUT&Tag, check the antibody affinity of the protein A or protein G that is fused with the Tn5. In general, proteins A and G have broad antibody affinity. However, protein A has a relatively higher affinity to rabbit antibodies, and protein G has a relatively higher affinity to mouse antibodies. Select the appropriate pA/pG-Tn5 that matches your secondary antibody. We recommend using mouse primary antibody, rabbit anti-mouse secondary IgG, and pA-Tn5 or using rabbit primary antibody, Guinea Pig anti-rabbit secondary IgG, and pG-Tn5. For nano-CUT&Tag, select the appropriate secondary nanobody type [9] that matches your IgG type of primary antibody. It is better to use validated primary antibodies in your CUT&Tag.
2. For plant nuclei isolation, do a pilot assay to test the effect of series concentration of Triton X-100 (e.g., 2 mL IsoB each with 0.1%, 0.25%, 0.5%, and 1% Triton X-100 for 100 mg ground tissues), and select the appropriate one for your plant tissues.
3. The nuclei can be collected by Con-A magnetic beads after Con-A binding (**steps 1–3** in Subheading 3.3) or by centrifugation at low speed (300×*g*). For nuclei input >100,000/reaction, the centrifuge method can be applied. After centrifugation, carefully aspirate the supernatant. It is better to leave ~10 μL of supernatant in the tube rather to avoid the loss of the nuclei. For small nuclei input, Con-A beads binding method is recommended.
4. Use different combinations of N50X (N501 to N508) and N70X (N701 to N712) with different barcodes for sample indexing.
5. For library construction, the first 72 °C for 3 min for extension *cannot* be skipped. Start from 12 to 14 PCR cycles in PCR reaction for H3K4me3 when using ~1 μg chromatin in each reaction. Overamplification of the library will lead to a high level of PCR duplicates in the NGS.
6. General recommendation for choosing regular CUT&Tag and its derivative methods: Use bulk-cell CUT&Tag for bulk histone modification profiling. When target chromatin protein has low abundance or with low number of DNA binding sites, B-CUT&Tag is recommended; if homemade nano-Tn5 is available, nano-CUT&Tag can be performed in bulk-cell and single mode (single histone modality) with more concise steps compared with regular CUT&Tag (Fig. 1). If one needs to detect multiple targets in a one reaction at the single-cell level,

the barcoded (multimodal) version of nano-CUT&Tag can be applied and accommodated to 10× Genomics system [7].

7. As the presence of multimodal barcoding sequences in ME-A adaptor (ME-A_bcd1, ME-A_bcd2, ME-A_bcd3), the custom sequencing primers are needed in NGS for the barcoded (multimodal) version of nano-CUT&Tag.

Acknowledgments

This work was financially supported by grants from the Key Research and Development Program of Zhejiang (2024SSYS0099) and the Project of Xianghu Laboratory (2023C1S02001).

Conflict of Interest The authors have no conflicts of interest to declare that are relevant to the content of this chapter.

References

1. Kaya-Okur HS, Wu SJ, Codomo CA et al (2019) CUT&Tag for efficient epigenomic profiling of small samples and single cells. Nat Commun 10: 1930. https://doi.org/10.1038/s41467-019-09982-5
2. Kaya-Okur HS, Janssens DH, Henikoff JG et al (2020) Efficient low-cost chromatin profiling with CUT&Tag. Nat Protoc 15:3264–3283. https://doi.org/10.1038/s41596-020-0373-x
3. Ouyang W, Zhang X, Peng Y et al (2021) Rapid and low-input profiling of histone marks in plants using nucleus CUT&Tag. Front Plant Sci 12:634679. https://doi.org/10.3389/fpls.2021.634679
4. Tao X, Feng S, Zhao T, Guan X (2020) Efficient chromatin profiling of H3K4me3 modification in cotton using CUT&Tag. Plant Methods 16: 120. https://doi.org/10.1186/s13007-020-00664-8
5. Tao X-Y, Guan X-Y, Hong G-J et al (2023) Biotinylated Tn5 transposase-mediated CUT&-Tag efficiently profiles transcription factor-DNA interactions in plants. Plant Biotechnol J 21: 1191–1205. https://doi.org/10.1111/pbi.14029
6. Stuart T, Hao S, Zhang B et al (2023) Nanobody-tethered transposition enables multifactorial chromatin profiling at single-cell resolution. Nat Biotechnol 41:806–812. https://doi.org/10.1038/s41587-022-01588-5
7. Bartosovic M, Kabbe M, Castelo-Branco G (2021) Single-cell CUT&Tag profiles histone modifications and transcription factors in complex tissues. Nat Biotechnol 39:825–835. https://doi.org/10.1038/s41587-021-00869-9
8. Deng Y, Bartosovic M, Kukanja P et al (2022) Spatial-CUT&Tag: spatially resolved chromatin modification profiling at the cellular level. Science 375:681–686. https://doi.org/10.1126/science.abg7216
9. Pleiner T, Bates M, Görlich D (2018) A toolbox of anti-mouse and anti-rabbit IgG secondary nanobodies. J Cell Biol 217:1143–1154. https://doi.org/10.1083/jcb.201709115

Chapter 4

Identification of Accessible Transcription Factor Binding Sites During the Seed-to-Seedling Transition

Benjamin J. M. Tremblay and Julia I. Qüesta

Abstract

Germination is a highly dynamic developmental transition from dry quiescent seeds to vegetative seedlings which involves major chromatin rearrangements and changes in transcriptional programs. Assay for transposase-accessible chromatin with sequencing (ATAC-seq) is a quick and reproducible method for quantifying relative chromatin accessibility genome-wide. In this chapter, we describe the use of ATAC-seq for profiling chromatin accessibility changes and differential transcription factor binding site accessibility during germination from enriched embryos.

Key words Seed germination, Embryos, Tagmentation, Chromatin accessibility, Epigenetics, Transcription factor binding sites, ATAC-seq, Nuclei

1 Introduction

A major developmental switch occurs during the seed-to-seedling transition when the mature embryo (dormant dry seed) develops into a vigorously growing seedling. This transition represents a change between two very distinct types of development: from embryonic (heterotrophic) growth to vegetative (autotrophic) growth, when the seedling gains photosynthetic competence. As such, the seed-to-seedling switch relies on complex cellular differentiation programs that must be tightly timed in response to both external environmental stimuli and endogenous signals within the plant. A near complete reprogramming of the plant transcriptome assures the completion of seed germination and early seedling development.

Transcriptomic analyses have revealed the distinct regulatory programs that are sequentially activated during the seed-to-seedling transition [1]. Expression of genes related to cellular catabolism, vesicle-mediated transport, seed development, and response to the phytohormone abscisic acid (ABA) are enriched in dry seeds,

José M. Franco-Zorrilla and Carmen Martín-Pizarro (eds.), *Plant Transcription Factors: Methods and Protocols*,
Methods in Molecular Biology, vol. 2985, https://doi.org/10.1007/978-1-0716-4972-5_4,

whereas cold stratification in the dark is characterized by RNA metabolic processes. Exposure to light triggers rapid cell elongation, requiring a higher rate of protein translation. The transition to post-germinative marks the start of DNA replication and cell division. Finally, a multitude of different seedling and growth pathways are activated, including photosynthesis, cell wall biogenesis, carbohydrate biosynthetic process, and root morphogenesis. This precise and timely sequence of events relies on an intricate network of transcription factors (TFs) that are induced downstream of the antagonistic activities of ABA and gibberellins (GA). Considerable efforts have been dedicated to building gene co-expression networks controlling seed development, germination, and early seedling growth, mostly based on transcriptomic studies [2–8]. However, TFs are usually expressed at low levels and may be subject to posttranslational modifications for activation, limiting the capacity of accurately predicting their downstream regulatory networks.

Chromatin regulation controls genome-wide accessibility of *cis*-regulatory elements (CREs), thus facilitating TF binding activity. CREs are noncoding sequences located proximally upstream (promoters) or distantly (enhancers and silencers) and harbor TF recognition motifs [9, 10]. Specific combinations of TFs binding to CREs determine the transcriptional outcome of target genes in response to differentiation, cell fate acquisition, and environmental stimuli. Mapping open chromatin regions allows the prediction of putative regulatory TFs controlling specific gene expression programs. Therefore, CRE identification combined with transcriptomic analysis has the potential to be a much more powerful approach to refine transcriptional networks driving developmental responses, which is particularly of interest in plant research where ChIP-seq datasets are still lacking.

The Assay for Transposase-Accessible Chromatin sequencing (ATAC-seq) has been adopted as the method of choice for profiling genome-wide chromatin accessible regions (ACRs) in plant and animal systems, both at the level of whole tissue and single cells [9–11]. ATAC-seq relies on the isolation of a relatively small number of nuclei which is particularly key when working with low input samples or complex tissues such as dry seeds. In this chapter, we describe a detailed protocol we used to perform embryo enrichment [12] and ATAC-seq during a time-course of *Arabidopsis* germination and early seedling establishment [1].

2 Materials

2.1 Embryo Enrichment

1. Sterilized *Arabidopsis thaliana* dry seeds, 200 mg per sample (*see* **Note 1**).

2. Sterile water.
3. Liquid ½ Murashige and Skoog (MS) media (*see* **Note 2**).
4. 40% sucrose made with sterile water, 6 mL per sample.
5. Cooled (4 °C) centrifuge with 15 mL tube adapters.
6. Microscope slides.
7. Miracloth.
8. Liquid nitrogen.
9. Mortar and pestle.

2.2 Nuclei Extraction

1. Nuclei extraction buffer 1 (NEB1): 20 mM MOPS, pH 7, 40 mM NaCl, 90 mM KCl, 2 mM EDTA, 0.5 mM EGTA, 0.5 mM spermidine, 0.2 mM spermine, protease inhibitor cocktail (e.g., Roche Complete EDTA-free protease inhibitor tablets) (*see* **Note 3**). 10.5 mL is needed per sample. Keep chilled.
2. Nuclei extraction buffer 2 (NEB2): 250 mM sucrose, 10 mM Tris-HCl, pH 8, 10 mM $MgCl_2$, 1% Triton X-100, protease inhibitor cocktail. 1.3 mL is needed per sample. Keep chilled.
3. Nuclei extraction buffer 3 (NEB3): 1.7 M sucrose, 10 mM Tris-HCl, pH 8, 2 mM $MgCl_2$, 0.15% Triton X-100, protease inhibitor cocktail. 300 μL is needed per sample. Keep chilled.
4. 100 μm and 40 μm cell strainers, one each per sample.
5. Cooled (4 °C) benchtop centrifuge with 15 mL tube adapters.

2.3 Counting of Nuclei

1. 0.2 μg/μL DAPI, 1 μL per sample.
2. Neubauer Improved chamber.
3. Epifluorescent microscope with mercury lamp, DAPI filter, and a 20× objective.

2.4 Tagmentation

1. Tagmentation master mix (1×): 2.5 μL loaded Tagmentase, 25 μL 2× Tagmentation buffer (Diagenode), 16.5 μL, PBS, pH 7.4, 0.5 μL 1% digitonin, 0.5 μL 10% Tween-20, 5 μL nuclease-free water (*see* **Note 4**).
2. Thermomixer (*see* **Note 5**).
3. MicroChIP DiaPure kit (Diagenode) (*see* **Note 6**).

2.5 Library Preparation

1. NEBNext High-Fidelity 2× PCR Master Mix (New England Biolabs), 30 μL per sample.
2. Nuclease-free water.
3. 10 μM Universal dual index (UDI) primers (e.g., Diagenode 24 UDI for Tagmented libraries Set I), 1.25 μL per sample.
4. 20× EvaGreen (Biotium), 1 μL per sample.

5. PCR thermocycler.
6. Quantitative PCR (qPCR) machine.

2.6 Library Purification and QC

1. Freshly prepared 80% ethanol.
2. AMPure XP or SPRIselect beads.
3. Magnetic rack.
4. Nuclease-free water.
5. Benchtop centrifuge.
6. Bioanalyzer 2100 or TapeStation with DNA High Sensitivity kit (Agilent).

2.7 Data Analysis

1. fastp (v0.23.4) [13].
2. bowtie2 (v2.4.4) [14].
3. samtools (v1.14) [15].
4. bedtools (v2.30.0) [16].
5. MACS3 (v3.0.0b3) [17].
6. IGV (v2.12.2) [18].
7. XSTREME (v5.4.1) [19].

3 Methods

3.1 Embryo Enrichment

1. Stratify *Arabidopsis thaliana* seeds in sterile water for 72 h at 4 °C in the dark using microcentrifuge tubes. Vortex well to evenly distribute seeds. Transfer seeds to 125 mL Erlenmeyer flasks containing 50 mL ½ MS media. Move flasks to a shaker set to 200 rpm in a growth chamber at 22 °C with a constant 100 μmol m^{-2} s^{-1} light intensity.
2. Collect seeds by passing the entirety of the liquid through a piece of Miracloth. Keep collected seeds on ice. Place a small amount of seeds on a microscope slide. Using a second microscope slide, gently push down on the seeds while rotating until the embryos are released (*see* **Note 7**). Using a small spatula, transfer the mix of embryo and seed coat into a small beaker containing 5 mL of ice-cold 40% sucrose kept on ice. Repeat until all seeds have been processed.
3. Using a pipette with a cut 1000 μL tip, transfer the entire volume to a 15 mL Falcon tube. Rinse the beaker with another 1 mL of ice-cold 40% sucrose, and transfer to 15 mL Falcon tube. Centrifuge at 3000×*g* for 10 min in a centrifuge cooled to 4 °C. Using a pipette with another cut 1000 μL tip, isolate the upper enriched embryo fraction by collecting onto a piece of Miracloth (*see* **Note 8**) (Fig. 1). Rinse with ice-cold sterile water, and then place the piece of Miracloth onto a paper

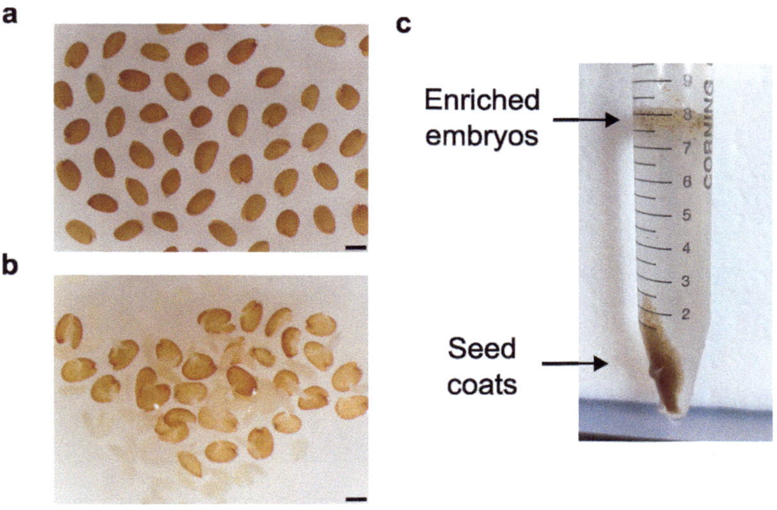

Fig. 1 Embryo enrichment from germinating *Arabidopsis thaliana* seeds. A small amount of seeds are first placed on a microscope slide (**a**) before being gently squeezed using a second slide, releasing the embryos without crushing them (**b**) (scale bar in the bottom right represents 1 mm). Centrifugation in 40% sucrose separates the seed coats, leading to an upper enriched embryo fraction (**c**)

towel to absorb excess water. Transfer enriched embryos to a mortar. Flash freeze with liquid nitrogen and grind to a powder with a pestle (*see* **Note 9**). The frozen ground tissue powder can be stored at −80 °C (*see* **Note 10**).

3.2 Nuclei Extraction

1. Add 10 mL ice-cold NEB1 directly to the mortar with the frozen ground powder and mix with the pestle. Collect with a 10 mL serological pipette, and filter using a 100 μm cell strainer into a 15 mL tube. Repeat using a 40 μm cell strainer.
2. Centrifuge at 1200×*g* for 10 min in a centrifuge cooled to 4 °C. Remove supernatant without disturbing the pellet. Resuspend in 1 mL ice-cold NEB2, and transfer into a 1.5 mL microcentrifuge using a pipette with a cut 1000 μL tip. Centrifuge at 12,000×*g* for 10 min in a centrifuge cooled to 4 °C. Remove supernatant without disturbing the pellet. Resuspend in 300 μL ice-cold NEB2.
3. Using a pipette with a cut 1000 μL tip, gently layer onto a 300 μL ice-cold NEB3 cushion in a new 1.5 mL microcentrifuge tube. Centrifuge at 16,000×*g* for 10 min in a centrifuge cooled to 4 °C. Remove supernatant and resuspend pellet in 500 μL NEB1. Transfer to a new 1.5 mL tube and keep on ice. Continue immediately with the rest of the protocol.

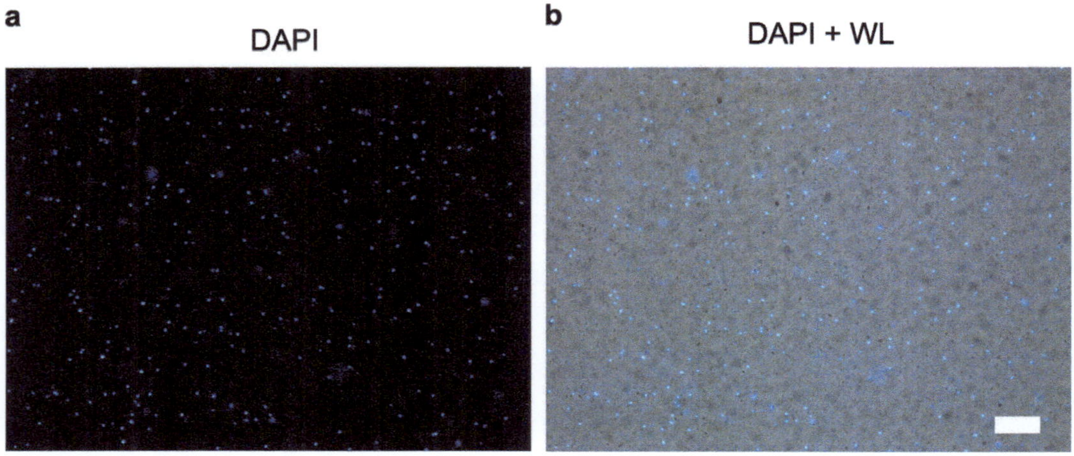

Fig. 2 DAPI-stained nuclei extracted from germinating seeds, shown with the blue channel only (**a**) and with both the blue and white light channels (**b**). Nuclei are shown inside a Neubauer Improved chamber using an epifluorescence microscope and a 20× objective (scale bar represents 200 μm)

3.3 Counting of Nuclei

1. Transfer 25 μL of the nuclei suspension to a new tube, and mix with 1 μL 0.2 μg/μL DAPI by pipetting. Keep on ice in the dark for 5 min.
2. Transfer 10 μL to a Neubauer Improved chamber and count nuclei in the central large square of the chamber using an epifluorescent microscope with a DAPI filter and a 20× objective (Fig. 2) (*see* **Note 11**). This value is equal to the number of nuclei per 0.1 μL.
3. Calculate the volume containing 50,000 nuclei. Transfer this amount to a new 1.5 mL tube.

3.4 Tagmentation and DNA Purification

1. Centrifuge at 1000×*g* for 10 min in a centrifuge cooled to 4 °C. Remove supernatant and resuspend nuclei in 50 μL tagmentation master mix. Incubate at 37 °C for 30 min in a thermomixer set to 900 RPM.
2. Add 250 μL Binding Buffer (Diagenode). Mix by pipetting. Transfer to a MicroChIP DiaPure column and centrifuge at 10,000×*g* for 30 s. Discard flow-through and add 200 μL wash buffer (Diagenode) to the column. Centrifuge as before, and then repeat the wash once. Transfer column to a 1.5 mL microcentrifuge tube. Add 12 μL nuclease-free water directly to the column matrix and centrifuge as before. Discard the column. The eluate can be stored at −20 °C.

3.5 Library Preparation

1. Prepare a PCR mix including 25 μL 2× NEBNext High Fidelity PCR Master Mix, 14 μL nuclease-free water, 10 μL tagmented DNA, and 1 μL 10 μM UDI primer mix. Place samples in a PCR thermocycler with the following program: [1] 72 °C for

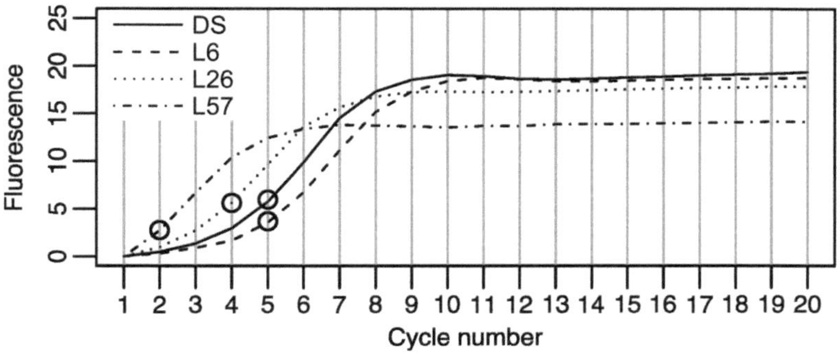

Fig. 3 qPCR fluorescence versus cycle number during library amplification from dry seed (DS), early germinating seed (L6), late germinating seed (L26), and seedling (L57) samples. The final cycle numbers before the reactions reach the linear phase are circled. These values are used for further amplification of the library PCR

5 min; [2] 98 °C for 30 s; [3] five cycles of 98 °C for 10 s; 63 °C for 30 s; and 72 °C for 1 min; [4] hold at 4 °C. Samples can be stored at −20 °C or kept at 4 °C for up to 2 h.

2. Prepare a qPCR mix including 5 μL of the previous PCR mix, 5 μL 2× NEBNext High Fidelity PCR Master Mix, 3.75 μL nuclease-free water, 0.25 μL 10 μM UDI primer mix, and 1 μL 20× EvaGreen. Place samples in a qPCR machine with the following program: [1] 98 °C for 30 s; [2] 20 cycles of 98 °C for 10 s; 63 °C for 30 s; 72 °C for 1 min, measuring fluorescence at the end of each cycle. For each sample, take note of the last cycle before the fluorescence reaches the linear phase (approximately one-third the maximum fluorescence) (Fig. 3).
3. Resume the PCR from before, setting the appropriate number of additional cycles for each sample as determined by qPCR. Use the following program: [1] 98 °C for 30 s; [2] cycles of 98 °C for 10 s; 63 °C for 30 s; 72 °C for 1 min; [3] hold at 4 °C. Samples can be stored at −20 °C.

3.6 Library Purification and QC

1. Allow Ampure XP beads to warm to room temperature (30 min). Vortex well and add 67.5 μL of beads with the previous PCR mix. Mix well and incubate for 5 min. Place on a magnetic rack for 5 min. Remove supernatant and add 200 μL 80% ethanol. Wait 30 s, remove liquid, and add 80% ethanol again. Wait 30 s and remove liquid. Let air-dry for up to 5 min without letting beads show signs of cracking. Remove from the magnetic rack and resuspend in 15 μL nuclease-free water. Incubate for 2 min, and then place on the magnetic rack. Once clear, transfer eluate to a new 1.5 mL tube and store at −20 °C (*see* **Note 12**).

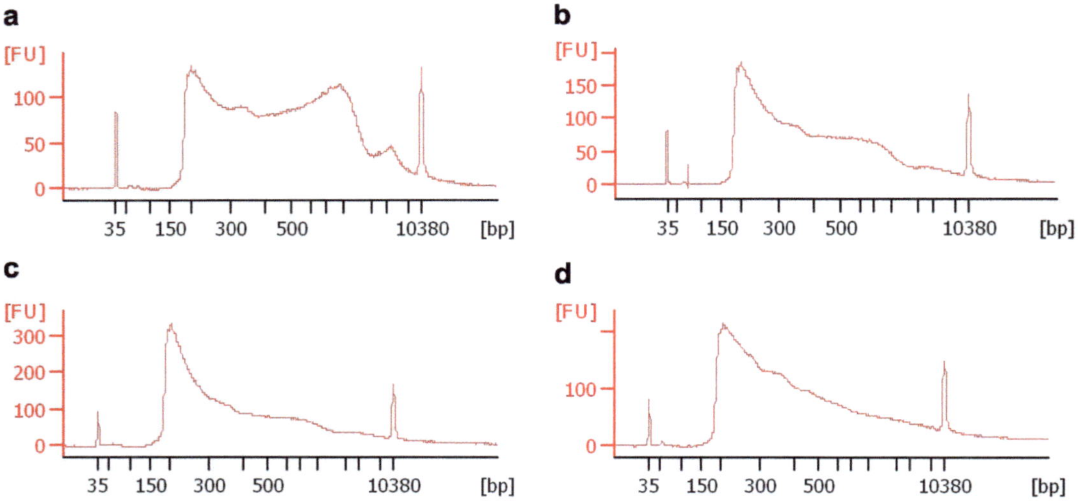

Fig. 4 Bioanalyzer 2100 profiles of ATAC-seq libraries using a DNA High Sensitivity chip from dry seeds (DS) (**a**), early germinating seeds (L6) (**b**), late germinating seeds (L26) (**c**), and seedlings (L57) (**d**). A clear peak corresponding to nucleosome-free regions can be observed around 200 bp, without any smaller fragments indicative of adapter contamination. Additional peaks corresponding to nucleosomal fragments may or may not be visible in the profiles of high quality libraries

2. Run 1 μL of the library on a DNA High Sensitivity chip for Bioanalyzer (Agilent). The resulting traces should show a peak at approximately 200 bp, with no visible smaller fragment peaks indicative of adapter contamination (Fig. 4). If these are present, repeat Ampure XP bead cleanup. Generally, library concentrations greater than 2 nM are sufficient for paired-end Illumina sequencing.

3.7 Data Analysis

We have found 30–40 million PE50 paired reads sufficient for differential motif enrichment analyses between conditions or genotypes.

1. Trim the FASTQ reads using *fastp -a CTGTCTCTTATACACATCT.*
2. Align reads to the *Arabidopsis thaliana* genome (TAIR10, downloaded from https://www.arabidopsis.org) using *bowtie2 --very-sensitive -X 2000.*
3. Use *samtools fixmate* to mark mate reads, *samtools sort* to sort reads, and *samtools markdup* to mark duplicates in the BAM.
4. Use *samtools view -F 3852 -q 30* to filter low quality reads, unmapped reads, alignments with unmapped mates, duplicates, secondary alignments, and supplementary alignments from the BAM. Several quality control checks can also be performed at this point, such as quantifying the fraction of duplicated and nonnuclear alignments (*see* **Note 13**).

5. Prepare a BED file containing the start and end coordinates of the five nuclear chromosomes, and then filter out chloroplast and mitochondrial alignments from the BAM using *samtools view* to keep those overlapping the BED regions.
6. Download an *Arabidopsis thaliana* blacklist (e.g., [20]) and remove overlapping alignments from the BAM using *bedtools intersect*.
7. Convert the BAM file to BED format using *bamtools bamtobed*.
8. Call peaks using *macs3 callpeak -f BED --nomodel --shift -100 --extsize 200 --keep-dup all -p 0.05 -g 1.191e8*.
9. Load the peak files into IGV to visually examine peak calling quality and decide on a *P*-value threshold. Merge peaks called from each sample using *bedtools merge* (*see* **Note 14**).
10. Use the *csaw* package to find differentially accessible peaks between conditions or genotypes (see the *csaw* book for further instructions: https://bioconductor.org/books/3.13/csawBook/). Obtain the DNA sequences of peaks of interest using *bedtools getfasta* before using XSTREME to find enriched transcription factor binding sites (*see* **Note 15**).

4 Notes

1. 200 mg of dry *Arabidopsis thaliana* seeds generally yield approximately 400 mg of enriched embryos sampled during imbibition and up to the transition point to post-germination, before cotyledon greening and seedling establishment.
2. We perform seed stratification using sterile water, making MS media unnecessary for such time-points.
3. Stocks of incomplete nuclei extraction buffers without spermidine, spermine, or protease inhibitor cocktail can be stored at 4 °C for several months.
4. Tagmentase and reaction buffers are available from other vendors. We have also successfully used homemade recombinant Tn5 transposase and buffers as described in [21].
5. A regular incubator can also be used by simply tapping tubes to resuspend nuclei every several minutes in place of constant shaking.
6. Alternate commercial cleanup kits such as MinElute PCR Purification Kit (Qiagen) or DNA Clean & Concentrator-5 (Zymo) can also be used, following the manufacturer's protocol.
7. Seeds must be squeezed between the microscope slides until the embryos are fully released from the seed coats in order for them to properly separate during centrifugation. Squeezing

too hard will crush the embryos and turn the surrounding liquid into an opaque beige color.

8. To further purify embryos, the upper embryo fraction can be mixed with additional 40% sucrose in a new tube and again centrifuged.
9. The enriched embryo fraction can also be chopped instead of flash frozen, which generally reduces background in the final ATAC-seq data. However, this requires immediately proceeding with nuclei extraction and transposition using only a single sample.
10. We typically perform embryo collection one sample at a time, and then perform nuclei extraction once four frozen ground samples are available.
11. This time can also be used to assess nuclei quality and the level of debris in the sample. High quality nuclei preps should consist of distinct, round nuclei with little debris.
12. The use of DNA LoBind tubes can help prevent loss of DNA during extended storage.
13. We have also created a tool to automate these filtering and quality control steps, which can be downloaded from https://github.com/bjmt/quaqc.
14. We generally decide which peaks to keep during merging on a per-experiment basis, depending on factors such as the number of samples overlapping peaks are present in and their peak *P*-values. An alternative approach is to keep all peaks which are highly reproducible between replicates as assessed by the Irreproducible Discovery Rate (IDR) method (see https://github.com/nboley/idr).
15. Single base-pair resolution TF footprints can also be analyzed using tools such as TOBIAS [22], which require the filtered BAM, peaks in BED format, and a list of motifs in JASPAR format (such as downloaded from https://jaspar.elixir.no). Motifs in alternate formats can be converted to JASPAR format using the *universalmotif* package (https://bioconductor.org/packages/universalmotif/).

Acknowledgments

We are grateful to Yajiao Cheng for providing an image of nuclei extracted from seeds. This work was supported by the Junior Leader Fellowship [LCF/BQ/PI19/11690003] from "laCaixa" Foundation [ID100010434] and by grants of the Spanish Ministry of Science and Innovation (MCIN/AEI/10.13039/501100011033), including PID2019-110510GA-I00 and

EUR2021-122003 awarded to J.I.Q. and the Severo Ochoa Excellence Programme for Centres (CEX2019-000902-S) awarded to CRAG. J.I.Q. is a Ramon y Cajal Fellow (RYC2021-032539-I). Work at CRAG was also supported by the Generalitat de Catalunya (AGAUR, GRE2021, Ref. SGR00873). B.J.M.T. holds an FPI predoctoral fellowship (PRE2019090156) from MICIN/AEI.

Conflict of Interest The authors have no conflicts of interest to declare that are relevant to the content of this chapter.

References

1. Tremblay BJM, Santini CP, Cheng Y et al (2024) Interplay between coding and non-coding regulation drives the Arabidopsis seed-to-seedling transition. Nat Commun 15: 1724. https://doi.org/10.1038/s41467-024-46082-5
2. Nakabayashi K, Okamoto M, Koshiba T et al (2005) Genome-wide profiling of stored mRNA in Arabidopsis thaliana seed germination: epigenetic and genetic regulation of transcription in seed. Plant J 41:697–709. https://doi.org/10.1111/j.1365-313X.2005.02337.x
3. Dekkers BJW, Pearce S, van Bolderen-Veldkamp RP et al (2013) Transcriptional dynamics of two seed compartments with opposing roles in Arabidopsis seed germination. Plant Physiol 163:205–215. https://doi.org/10.1104/pp.113.223511
4. Narsai R, Gouil Q, Secco D et al (2017) Extensive transcriptomic and epigenomic remodelling occurs during Arabidopsis thaliana germination. Genome Biol 18:172. https://doi.org/10.1186/s13059-017-1302-3
5. Okamoto M, Tatematsu K, Matsui A et al (2010) Genome-wide analysis of endogenous abscisic acid-mediated transcription in dry and imbibed seeds of Arabidopsis using tiling arrays. Plant J 62:39–51. https://doi.org/10.1111/j.1365-313X.2010.04135.x
6. Kawakatsu T, Nery JR, Castanon R, Ecker JR (2017) Dynamic DNA methylation reconfiguration during seed development and germination. Genome Biol 18:171. https://doi.org/10.1186/s13059-017-1251-x
7. Narsai R, Law SR, Carrie C et al (2011) In-depth temporal transcriptome profiling reveals a crucial developmental switch with roles for RNA processing and organelle metabolism that are essential for germination in Arabidopsis. Plant Physiol 157:1342–1362. https://doi.org/10.1104/pp.111.183129
8. Silva AT, Ribone PA, Chan RL et al (2016) A predictive coexpression network identifies novel genes controlling the seed-to-seedling phase transition in Arabidopsis thaliana. Plant Physiol 170:2218–2231. https://doi.org/10.1104/pp.15.01704
9. Marand AP, Eveland AL, Kaufmann K, Springer NM (2023) cis-regulatory elements in plant development, adaptation, and evolution. 74:111–137. https://doi.org/10.1146/ANNUREV-ARPLANT-070122-030236
10. Schmitz RJ, Grotewold E, Stam M (2022) Cis-regulatory sequences in plants: their importance, discovery, and future challenges. Plant Cell 34:718–741. https://doi.org/10.1093/plcell/koab281
11. Klemm SL, Shipony Z, Greenleaf WJ (2019) Chromatin accessibility and the regulatory epigenome. Nat Rev Genet 20:207–220. https://doi.org/10.1038/s41576-018-0089-8
12. Iwasaki M, Lopez-Molina L (2021) Arabidopsis thaliana mature endosperm dissection and isolation of genomic DNA from mature seed tissues. Methods Mol Biol 2250:239–243. https://doi.org/10.1007/978-1-0716-1134-0_22
13. Chen S, Zhou Y, Chen Y, Gu J (2018) fastp: an ultra-fast all-in-one FASTQ preprocessor. Bioinformatics 34:i884–i890. https://doi.org/10.1093/bioinformatics/bty560
14. Langmead B, Salzberg SL (2012) Fast gapped-read alignment with Bowtie 2. Nat Methods 9: 357–359. https://doi.org/10.1038/nmeth.1923
15. Li H, Handsaker B, Wysoker A et al (2009) The sequence alignment/map format and SAMtools. Bioinformatics 25. https://doi.org/10.1093/bioinformatics/btp352
16. Quinlan AR, Hall IM (2010) BEDTools: a flexible suite of utilities for comparing genomic

features. Bioinformatics 26. https://doi.org/10.1093/bioinformatics/btq033

17. Zhang Y, Liu T, Meyer CA et al (2008) Model-based analysis of ChIP-Seq (MACS). Genome Biol 9. https://doi.org/10.1186/gb-2008-9-9-r137
18. Robinson JT, Thorvaldsdóttir H, Winckler W et al (2011) Integrative genomics viewer. Nat Biotechnol 29:24–26
19. Grant CE, Bailey TL (2021) XSTREME: comprehensive motif analysis of biological sequence datasets. bioRxiv. https://doi.org/10.1101/2021.09.02.458722
20. Klasfeld S, Roulé T, Wagner D (2022) Greenscreen: a simple method to remove artifactual signals and enrich for true peaks in genomic datasets including ChIP-seq data. Plant Cell 34:4795–4815. https://doi.org/10.1093/plcell/koac282
21. Najle SR, Grau-Bové X, Elek A et al (2023) Stepwise emergence of the neuronal gene expression program in early animal evolution. Cell 186:4676–4693.e29. https://doi.org/10.1016/j.cell.2023.08.027
22. Bentsen M, Goymann P, Schultheis H et al (2020) ATAC-seq footprinting unravels kinetics of transcription factor binding during zygotic genome activation. Nat Commun 11:4267. https://doi.org/10.1038/s41467-020-18035-1

Part II

Visualization and Quantification of TF-DNA and TF-Protein Interactions

Chapter 5

Dual-Luciferase Assay System for the Quantification of Promoter Activity

Patricia Ballester and Cristina Ferrándiz

Abstract

Transient luciferase reporter assays are a widely used and powerful tool for investigating promoter activity and gene regulation in plant systems. These assays make use of the enzymatic conversion of a luciferin substrate into bioluminescence to provide a highly sensitive and reproducible measure of transcriptional activity. The dual-luciferase system, incorporating Firefly luciferase (LUC) as the primary reporter and Renilla luciferase (REN) as an internal control, enhances experimental accuracy by normalizing for variations in infiltration efficiency, tissue viability, and environmental conditions. Agroinfiltration-mediated transient expression in *Nicotiana benthamiana* leaves enables rapid and high-throughput analysis of transcriptional regulation, including the functional characterization of transcription factors (TFs), *cis*-regulatory elements, and enhancer/silencer activity. This approach facilitates systematic promoter dissection through site-directed mutagenesis, deletion mapping, and combinatorial TF studies to uncover complex regulatory interactions. Here, we present a detailed protocol for performing dual-luciferase transient expression assays, emphasizing best practices for ensuring robust and reproducible results.

Key words Luciferase, Transcription factor (TF), Promoter, Agroinfiltration

1 Introduction

Transient expression of luciferase reporter genes in plant tissues is a powerful and versatile method to study promoter activity and gene regulation. Luciferase-based assays provide a highly sensitive and reproducible means to measure transcriptional activity. The principle of these assays relies on the enzymatic conversion of a luciferin substrate into light by the luciferase enzyme. The reporter gene encoding luciferase is placed under the control of a promoter-of-interest and subsequently introduced into plant cells, typically via agroinfiltration—a transient expression technique that facilitates high-throughput functional studies. The emitted bioluminescence is then detected using a luminometer or imaging system, with signal intensity directly correlating with promoter-driven transcriptional activity.

José M. Franco-Zorrilla and Carmen Martín-Pizarro (eds.), *Plant Transcription Factors: Methods and Protocols*, Methods in Molecular Biology, vol. 2985, https://doi.org/10.1007/978-1-0716-4972-5_5,

To improve experimental accuracy and control for variations in infiltration efficiency, tissue viability, and environmental conditions, the dual-luciferase system is commonly employed. In this system, Firefly luciferase (LUC) serves as the primary reporter, while Renilla luciferase (REN) acts as an internal control [1]. The ratio of LUC to REN activity allows for normalization, reducing technical variability and increasing the reliability of the results.

One of the major advantages of this system is its flexibility in studying gene regulatory mechanisms. Unlike stable transformation methods, which require extensive time and labor, transient expression allows for rapid testing of promoter activity, the influence of transcription factors (TFs), and the functional characterization of *cis*-regulatory elements. A frequent application of this assay involves analyzing the effects of TFs on promoter activation or repression by co-infiltrating luciferase reporter constructs alongside effector TFs. If a TF activates the promoter under investigation, luciferase activity will be elevated compared to a control where only the reporter construct is introduced. Conversely, if the TF represses gene expression, the luminescence signal will be lower than the control.

Beyond single-factor regulation, this system can be adapted to study the interactions between multiple TFs on the same promoter. By co-infiltrating different TFs simultaneously, researchers can explore synergistic, antagonistic, or combinatorial effects on transcriptional regulation, providing deeper insights into complex regulatory networks.

Furthermore, this approach allows for precise dissection of promoter architecture by targeting specific regions for functional analysis. For instance, site-directed mutagenesis can be employed to alter putative *cis*-regulatory elements, thereby assessing their contributions to transcriptional activity. Similarly, serial deletion constructs can be generated to systematically map key regulatory sequences within a promoter region, enabling the identification of essential motifs required for gene expression control.

In addition to promoters, the luciferase assay system is a valuable tool for investigating other regulatory elements, such as enhancers and silencers. These elements, which modulate gene expression in a distance- and orientation-independent manner, can be located upstream, downstream, or within introns of a gene. To assess their functional significance, putative regulatory sequences can be cloned upstream of a minimal promoter-luciferase construct, allowing for direct evaluation of their ability to enhance or suppress transcription.

Overall, transient luciferase reporter assays provide a robust and efficient platform for deciphering plant gene regulation. Their high sensitivity, rapid execution, and adaptability make them essential for exploring transcriptional mechanisms, identifying regulatory elements, and studying complex gene regulatory networks in plant systems.

Here we describe the protocol we use routinely, mediated by agroinfiltration of *Nicotiana benthamiana* leaves of dual-luciferase reporters.

2 Materials

2.1 Agroinfiltration of Nicotiana benthamiana Leaves

- Competent *Agrobacterium tumefaciens* strain C58 containing the vectors for the assay.
  - pGreen binary vector containing the promoter-of-interest upstream of the fLUC reporter (Fig. 1).
  - Binary vectors containing the TF effectors under a CaMV35S or other constitutive promoter.
  - p19 helper plasmid.
  - A CaMV35S:fLUC reporter that can serve as positive control.
- Luria-Bertani (LB) agar plates and liquid medium supplemented with the corresponding antibiotics.
- Incubator chamber at 28 °C.
- Inoculation sterile loops.

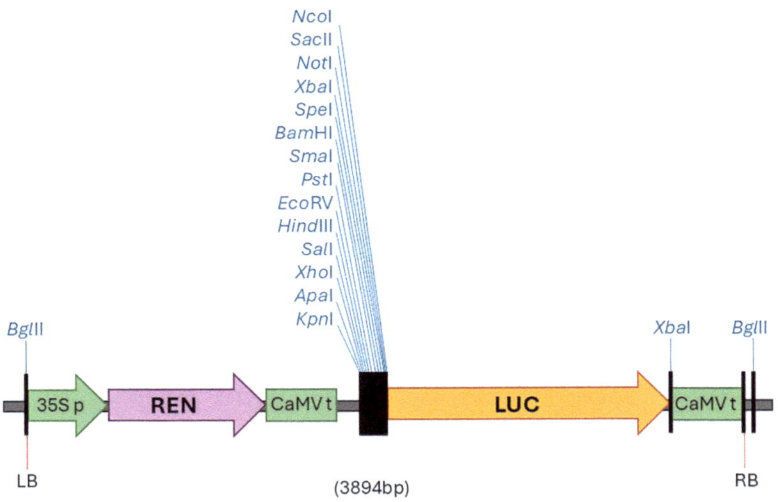

Fig. 1 T-DNA representation of the pGreenII 0800-LUC vector used for transient expression. The 3894-bp T-DNA, from the left border (LB) to the right border (RB), includes the *RENILLA* gene, represented as a purple arrow, and the *LUCIFERASE* gene, as a yellow arrow. The 35S promoter and CaMV terminator are indicated with a green arrow and a green square, respectively. The restriction sites are shown in blue

- 250 mL Erlenmeyer flask.
- Shaking incubator at 28 °C.
- 50 mL conical Falcon tubes.
- Benchtop and swing bucket centrifuge.
- 0.1 M MES stock solution, pH 5.6 adjusted with NaOH. Sterilized by autoclaving.
- 100 mM acetosyringone stock solution in DMSO; store at −20 °C. Sterilized by filtration.
- 1 M $MgCl_2$ stock solution. Sterilized by filtration.
- Orbital shaker.
- Spectrophotometer and matching cuvettes.
- 2 mL disposable sterile syringes.
- Growth chamber under long day conditions (16-h light, 8-h dark) at 22 °C.
- *Nicotiana benthamiana* plants, approximately 4 weeks old.

2.2 Luciferase Activity Measurement

- Dual-Glo® Luciferase Assay System kit containing.
  - Luciferase Assay Buffer II (LARII) with dissolved lyophilized Luciferase Assay Substrate. Always keep in the dark and store at −20 °C before preparation (*see* **Note 1**).
  - Stop & Glo® Reagent with Stop & Glo® Substrate and Stop & Glo® Buffer. Always keep in the dark and store at −20 °C.
- 1.3 cm diameter cork-borer.
- 1.5 mL Eppendorf tubes.
- Liquid nitrogen.
- Tissue homogenizers pestles, or TissueLyser.
- Glass beads for TissueLyser.
- Passive Lysis 5× Buffer (Promega).
- Vortex.
- Ice.
- Table-top centrifuge for 1.5 mL Eppendorf tubes.
- 96-well luminometry plates.
- Luminometer *GLOMAX MULTI DETECTION SYSTEM—PROMEGA (MODEL:9301-062).*

3 Methods

3.1 Generation of LUC Reporter Constructs

Here we provide details for generating constructs in the pGreen::LUC vector described in [2]. Alternatively, it is possible to use modular systems based on Golden Gate [3] or Golden Braid cloning [4–6] to reduce the number of constructs that need to be transiently expressed (*see* **Note 2**).

1. Analysis of promoter proximal region.

   Promoter regions of genes of interest must be amplified from genomic DNA with primer pairs including restriction enzymes sites present in the pGreen::LUC vector cloning site (*see* **Note 3**).

   Usually, the amplified promoter regions are cloned into vectors suitable for PCR products, digested with selected restriction enzymes, and ligated into pGreen::LUC [2] to generate the reporters for transient expression assays in *Nicotiana benthamiana* leaves (*see* **Note 4**).

2. Analysis of other regions of interest.

   When testing specific domains within a promoter, intron, or other genomic regions with putative regulatory function (or likely bound by the TF of interest), these must be cloned upstream of a minimal 35S promoter and the LUC reporter (*see* **Notes 5** and **6**).

   When the tested domain contains a putative binding site for the TF of interest, it is also possible to generate a similar construct but mutating the identified binding site to further confirm that the TF under study effectively binds in that region.

   If there is no information available about the putative domains where the TF binds to the target promoter under study, it is possible to check different fragments of the promoter (always adding the minimal 35S promoter before the LUC CDS) to generate a map of the regions that contain the binding sites.

3.2 Cloning of Effectors

The coding region of the TF of interest is usually cloned under the control of the CaMV 35S promoter or any other constitutive promoter, in any of the commonly used expression vectors for plants.

For some experimental designs, it may be useful to modify the activity of the TF to be tested in order to make them constitutive transcriptional activators or repressors. For this purpose, the coding region of the VP16 activator domain [7, 8] or the EAR repression motif [9] can be cloned in phase with the CDS of the TF coding region. By using this type of effectors, it is possible to differentiate if a given TF does not act on a promoter region because it does not bind or it binds but it is not able to activate or repress the transcription of the LUC gene. Furthermore, if you do not have a basal activity of your promoter that allows testing for repressor activity of a given effector, you can solve the problem with these domains.

3.3 Agroinfiltration of *Nicotiana benthamiana* Leaves

1. Three days before infiltration: Refresh the *Agrobacterium* strains from plate or glycerol stocks by triple streak in LB medium supplemented with the corresponding antibiotics, and incubate at 28 °C for 2–3 days.

   For each assay, in addition to the *Agrobacterium* with the promoter:LUC construct and those with the TFs under study, two additional strains have to be refreshed, containing 35S::LUC and pGreen::LUC (empty vector) for extra positive and negative controls and another one with the viral suppressor of posttranscriptional gene silencing p19.
2. One day before infiltration: Inoculate a colony of each *Agrobacterium* in 50 mL of liquid LB medium supplemented with antibiotics (in 250 mL flasks). Incubate at 28 °C with shaking for approximately 24 h (to saturation).
3. On infiltration day: Transfer the *Agrobacterium* cultures to centrifuge tubes, and collect the cells by centrifugation at 1700 × g for 30 min at room temperature.
4. Resuspend the pellets with 10 mL of infiltration solution (10 mM $MgCl_2$; 10 mM MES; 200 μM acetosyringone in MilliQ H_2O).
5. Incubate the cultures, placing the tubes horizontally, in darkness, for 90 min, with shaking (200 rpm) at 28 °C.
6. Measure the optical density (OD) of the culture and adjust to OD ~ 0.4 in a final volume of 10 mL with infiltration solution in 50 mL Falcon tubes.
7. Incubate the cultures, placing the tubes horizontally, in darkness, for 90 min, with shaking (50 rpm) at 28 °C.
8. Prepare in Falcon tubes the mixes to infiltrate *Nicotiana benthamiana* leaves in a final volume of 10 mL (enough to infiltrate four leaves). These mixes will contain the *Agrobacterium* with the expression of luciferase directed by the promoter to be tested, the *Agrobacterium* with the TF whose binding to the promoter we want to test, and the *Agrobacterium* carrying the p19 vector (*see* **Notes 7** and **8**).
9. Before infiltration, it is advisable to moisten the leaves with water to promote subsequent infiltration (Fig. 2).
10. Infiltrate non-damaged leaves from healthy *Nicotiana benthamiana* plants. You must infiltrate four leaves of different plants per assay with the aid of a 2 mL syringe without needle. The infiltration is done placing the filled syringe on the abaxial side of the leaf and pressing the plunger while exerting a pressure with one finger on the other side of the leaf (*see* **Note 9**).
11. Place the infiltrated plants back into a controlled growth chamber for 3 days.

Fig. 2 Procedure of agroinfiltration of *Nicotiana benthamiana* leaves and sampling. (**a**) Healthy *Nicotiana benthamiana* plant at the optimal stage of agroinfiltration. (**b**) Infiltration with a syringe on the abaxial side of the leaf. (**c**, **d**) Obtainment of the samples using a cork-borer. (**e**, **f**) Placement of the samples without causing any damage in an Eppendorf tube

3.4 Luciferase Activity Measurement

1. We typically use the Dual-Glo® Luciferase Assay System kit. The first step is to calculate the volume for all the reagents required to perform the assay: "Passive Lysis Buffer (PLB)," "Luciferase Assay Reagent II (LAR-II)," and "Stop & Glo Reagent" (*see* **Notes 10** and **11**).
2. Obtain the samples using a cork-borer of approximately 1.3–1.5 cm in diameter. You must obtain one cylinder per infiltrated leaf, place the samples in a 1.5 mL Eppendorf tube and freeze immediately in liquid nitrogen. Keep them in a −80° freezer until use (*see* **Notes 12** and **13**).
3. Grind the plant material in the presence of liquid nitrogen until a fine powder is obtained (*see* **Note 14**).
4. Add 350 μL of PLB 1×.
5. Centrifuge the samples for 5 min at 4 °C.
6. Transfer 50 μL of the supernatant to the microplate to quantify luminescence.
7. Add 40 μL of LAR-II to each sample, and mix carefully, avoiding to generate bubbles in the wells.
8. Incubate for 10 min at room temperature.

9. Start reading the luciferase activity with the following parameters: integration time, 10 s; period, 2 min and three readings (*see* **Note 15**).
10. Add 40 μL "Stop & Glo Reagent" to each sample, and mix carefully, avoiding to generate bubbles in the wells.
11. Incubate for 10 min at room temperature.
12. Start reading the Renilla activity with the following parameters: integration time, 10 s; period, 2 min and three readings.
13. Save the data and transfer them as an Excel data on your own computer (*see* **Notes 16** and **17**).
14. Calculate the relative ratio of luciferase/Renilla activity to normalize variations caused by different transfection efficiency.
15. Obtain the average of the relative ratios from the three technical replicates for each interaction you want to test and compare it with the promoter activity without effectors.
16. Calculate and normalize standard deviation.
17. Repeat the experiment three times to generate three independent biological replicates with three technical replicates per sample.

4 Notes

1. Prepare the LAR II buffer by adding 1 g of Luciferase Assay Substrate to the 10 mL of LAR II buffer and store at −80 °C. You must aliquots this solution because it is degraded by thawing, and never use it more than two times.
2. GoldenBraid cloning is a standardized DNA assembling system designed to generate a multipartite assembly of basic DNA parts, which are then incorporated to a double loop cloning design that allows binary assembly of multipartite constructs [4, 10]. One of the biggest advantages of using this system in the luciferase assay is that you can combine in the same construct several transcriptional units, such as the promoter:LUC, the 35S:REN, and the p19 or, more importantly, several putative effectors (e.g., if synergic or antagonistic interactions of different factors on the same promoter are explored), minimizing the number of *Agrobacterium* that must infect a cell and ensuring that all transformed cells receive the same dose of each effector.
3. When testing a full promoter region, we recommend including the full 5′UTR region and adding an NcoI site in the 3′ amplification primer, since it will use the ATG of the restriction site (mimicking that of the tested gene) as the start codon of the LUC gene when cloned in the pGreen::LUC vector.

4. In our hands, the maximum length of a promoter that we have been able to test successfully is 6 kb.
5. You can fuse the region of interest and the 35S minimal promoter by PCR designing partially overlapping primers that include also the restriction sites necessary to clone the resulting fragment upstream of LUC reporter.
6. Try using promoter fragments not shorter than 100 bp.
7. Prepare the *Agrobacterium* culture combinations from cultures at DO 0.4, diluting them 1/10 for the effectors, 1/50 for the reporters, and 1/50 for p19 with fresh infiltration solution (e.g., 200 μL of reporter, 200 μL of p19, and 1 mL of effectors in a total volume of 10 mL).
8. If you want to analyze the ability of a heterodimer to bind the promoter, add both effectors in the same culture combination.
9. Ideally, *Nicotiana* plants will have grown for 3–4 weeks under long days, presenting approximately five leaves.
10. Prepare the dilution 1× PLB from stock solution at 5× with distilled water and the dilution 1× "Stop & Glo Reagent" mixing the 100× Stop & Glo Substrate and Stop & Glo Buffer.
11. Preferably, thaw the aliquots of LAR II buffer you need for the assay in a water bath at room temperature and in dark.
12. Try to obtain samples from a healthy surface without wounds and minimize mechanical damage in the process.
13. Clean the cork-borer with ethanol between different samples.
14. If you proceed with the disruption of the material with a pestle, or with a TissueLyser, grind thoroughly, but do not allow the tissue to thaw and proceed immediately with the next step.
15. It is better to start the reading with the negative controls and leave the positive controls for the end, since high levels of luminescence may affect subsequent readings in adjacent wells.
16. Do not open the lid of the luminometer while reading.
17. You can reuse the plate. To do so, immediately clean the plate with extran, rinse with distilled water, and finally, extra rinse with absolute ethanol. Leave the plate to completely dry before storing it, cover with paper, and store it in dark.

Acknowledgments

This work was supported by grant CIPROM/2022/1 from Generalitat Valenciana to C.F.

Conflict of Interest The authors have no conflicts of interest to declare that are relevant to the content of this chapter.

References

1. Sherf BA, Navarro SL, Hannah RR, Wood KV (1996) Dual-luciferase TM reporter assay: an advanced co-reporter technology integrating firefly and renilla luciferase assays
2. Hellens RP, Allan AC, Friel EN et al (2005) Transient expression vectors for functional genomics, quantification of promoter activity and RNA silencing in plants. Plant Methods 1: 13. https://doi.org/10.1186/1746-4811-1-13
3. Engler C, Gruetzner R, Kandzia R, Marillonnet S (2009) Golden gate shuffling: a one-pot DNA shuffling method based on type IIs restriction enzymes. PLoS One 4:e5553. https://doi.org/10.1371/journal.pone.0005553
4. Sarrion-Perdigones A, Palaci J, Granell A, Orzaez D (2014) Design and construction of multigenic constructs for plant biotechnology using the GoldenBraid cloning strategy. Methods Mol Biol 1116:133–151. https://doi.org/10.1007/978-1-62703-764-8_10
5. Sarrion-Perdigones A, Vazquez-Vilar M, Palací J et al (2013) GoldenBraid 2.0: a comprehensive DNA assembly framework for plant synthetic biology. Plant Physiol 162:1618–1631. https://doi.org/10.1104/pp.113.217661
6. Sarrion-Perdigones A, Falconi EE, Zandalinas SI et al (2011) GoldenBraid: an iterative cloning system for standardized assembly of reusable genetic modules. PLoS One 6:e21622. https://doi.org/10.1371/journal.pone.0021622
7. Parcy F, Nilsson O, Busch MA et al (1998) A genetic framework for floral patterning. Nature 395:561–566. https://doi.org/10.1038/26903
8. Ng M, Yanofsky MF (2001) Activation of the Arabidopsis B class homeotic genes by APETALA1. Plant Cell 13:739–753. https://doi.org/10.1105/tpc.13.4.739
9. Hiratsu K, Matsui K, Koyama T, Ohme-Takagi M (2003) Dominant repression of target genes by chimeric repressors that include the EAR motif, a repression domain, in Arabidopsis. Plant J 34:733–739. https://doi.org/10.1046/j.1365-313x.2003.01759.x
10. Vazquez-Vilar M, Quijano-Rubio A, Fernandez-Del-Carmen A et al (2017) GB3.0: a platform for plant bio-design that connects functional DNA elements with associated biological data. Nucleic Acids Res 45: 2196–2209. https://doi.org/10.1093/NAR/GKW1326

Chapter 6

Yeast One-Hybrid (Y1H) Assay for Single Promoter-Transcription Factor Interaction in Higher Plants

Aileen Turner, Paz E. Zuñiga, and Carlos R. Figueroa

Abstract

The yeast one-hybrid (Y1H) assay is a powerful molecular biology tool for studying protein-DNA interactions. This technique involves using a single DNA bait cloned upstream of a reporter gene in a yeast plasmid to determine whether a specific transcription factor (TF) or other DNA-binding protein (referred to as "prey") can bind to the DNA bait. Typically, this technique employs a complete cDNA library to obtain multiple prey proteins, which are then introduced into yeast cells containing the bait construct. In this protocol, we present a simplified method not centered on finding new prey from cDNA libraries but on analyzing specific target plant proteins. The protocol provides a step-by-step guide for cloning DNA-bait and protein-prey constructs, transforming yeast, and screening for reporter interactions.

Key words Yeast one-hybrid, Promoter, Transcription factor, Protein-DNA interactions

1 Introduction

Dissecting complex interactions between proteins and DNA has significantly advanced our understanding of critical roles played by promoters and transcription factors (TFs) in plant gene regulation and cellular functions [1]. The techniques used to unravel these TF-DNA interactions can be classified as "TF-centered" or "gene-centered" methods [2]. "TF-centered" method focused on finding which regions of DNA bind to a specific TF employing techniques such as chromatin immunoprecipitation (ChIP) [3–6], DNA affinity purification and sequencing (DAP-seq) [7, 8], systematic evolution of ligands by exponential enrichment (SELEX) [9], or protein-binding microarrays [10–13]. Conversely, "gene-centered" methods aim to identify which TFs bind to a specific DNA sequence. Among the "gene-centered" approaches, since its development in the early 1990s, the yeast one-hybrid (Y1H) assay has become an important technique for detecting protein-DNA interactions (PDIs) [2, 14].

José M. Franco-Zorrilla and Carmen Martín-Pizarro (eds.), *Plant Transcription Factors: Methods and Protocols*,
Methods in Molecular Biology, vol. 2985, https://doi.org/10.1007/978-1-0716-4972-5_6,

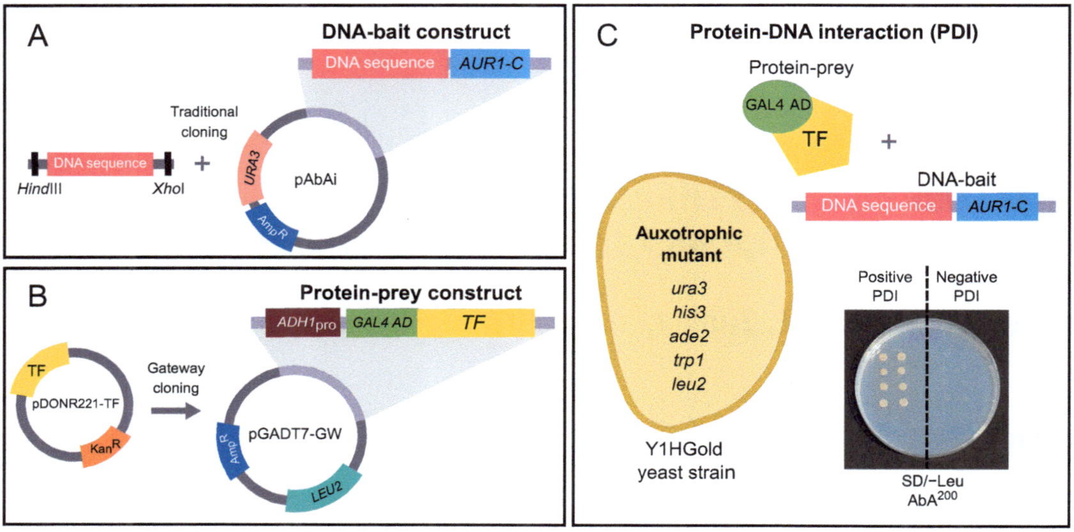

Fig. 1 Schematic representation of yeast one-hybrid (Y1H) assay for detecting physical PDIs in higher plants, including (**a**) DNA-bait cloning, (**b**) protein-prey construct cloning, and (**c**) protein-DNA interaction (PDI) on Y1HGold yeast strain. The presence of yeast colonies in droplets on restricted media (SD/-Leu AbA^{200}) indicates a positive PDI, while the absence of colony growth signifies a negative PDI. AbA, aureobasidin A; Amp^R, ampicillin resistance; $ADH1_{pro}$, alcohol dehydrogenase 1 promoter; *AUR1-C*, inositol phosphoryl ceramide synthase (IPCS) mutant gene; *GAL4 AD*, galactose-responsive transcription factor activation domain sequence; *LEU2*, beta-isopropylmalate dehydrogenase gene; Kan^R, kanamycin resistance; TF, specific plant transcription factor DNA sequence; *URA3*, orotidine 5-phosphate decarboxylase gene

The basic yeast one-hybrid (Y1H) assay is a method for detecting physical PDIs that includes three basic components: [1] a reporter construct (DNA-bait) consisting of a DNA sequence such as a promoter, an enhancer region of a gene or an independent *cis*-regulatory element cloned upstream of a reporter gene; [2] a protein-prey construct for the expression of a specific TF or other DNA-binding protein fused to a yeast galactose-responsive transcription factor activation domain (GAL4 AD); and [3] a specific yeast strain used for Y1H, harboring mutations in multiple auxotrophic genes that make them incapable of growing in the absence of some amino acids and nutrients in the media [2] (Fig. 1). The physical PDI between the protein-prey and DNA-bait activates the reporter gene expression in yeast and allows the interaction to be screened. Heterologous positive interactions in yeast suggest PDI in vivo.

The system is based on the ability of yeast to integrate foreign DNA-baits into their genome by homologous recombination and express the protein-prey construct AD-TF fusion proteins to interact with them, activating a reporter gene. To ensure the presence of DNA-baits and protein-prey components of the system and avoid false negatives, specific yeast strains are used in Y1H that usually lack functional copies of genes important for amino acid or nutrient syntheses like *ura3* (uracil), *leu2* (leucine), *trp1* (tryptophan), and *his3* (histidine) and prevent them from growing in nutrient restricted media [14]. The function of these genes can then be replaced by functional copies in the yeast plasmids used to clone the DNA-bait and protein-prey construction, allowing yeast to grow if they are present.

For example, in this method, we employ a Y1HGold yeast strain from Clontech Matchmaker® Gold Yeast One-Hybrid Library Screening System, which has been modified to require exogenous uracil and leucine for growth. DNA-baits are generated by cloning a target DNA into the pAbAi yeast plasmid, rescuing the function of the *URA3* gene when it is integrated into the yeast genome. Moreover, the protein-prey construction is cloned by Gateway Technology® into a pGADT7 plasmid that includes a functional copy of the *LEU2* gene. Each plasmid allows yeast to grow in specific restricted media.

Different PDI screening systems have been used in Y1H, including the colorimetric detection of β-Gal and *lacZ* activity, expression of the auxotrophic gene *HIS3*, and antibiotic resistances, which are also used in yeast two-hybrid (Y2H) assays. Aureobasidin A (AbA) is a common antibiotic used in Y1H, which inhibits the inositol phosphoryl ceramide synthase (IPCS) enzyme and is toxic to yeast at low concentrations (0.1–0.5 μg/mL) [15, 16]. For screening for PDIs, a mutant *IPCS* gene (*AUR1-C*) could be used as a reported gene and fusion with a DNA fragment of interest to make a reported construct (DNA-bait) [17].

Like any laboratory technique, Y1H has a few limitations that must be highlighted. These limitations can lead to missed true interactions or identify PDI that do not occur in vivo [18]. Firstly, Y1H is restricted to using yeast as a heterologous expression system for analyzing protein-DNA interactions, which may only partially replicate the regulatory mechanisms found in higher plants. Additionally, the cellular machinery in yeast may be insufficient for the correct expression of some proteins [19, 20]. This can lead to improper posttranscriptional or posttranslational processing, incorrect protein folding, or lack of necessary posttranslational modifications, all of which can impact the function and binding abilities of the proteins [21, 22]. Therefore, PDI identified using Y1H must include specific and appropriate controls and additional validation with other methods, such as a luciferase reporter assay [23]. This recommendation is also applicable to all different PDI detection methods.

Also, Y1H assays only report physical interactions between a protein and DNA, implying that yeast AD activates the reporter regardless of the TF's function and cannot distinguish between a TF acting as an activator or repressor.

Nevertheless, these limitations aside, Y1H has been used to elucidate several PDIs in higher plants, including rice (*Oryza sativa*) [24], commercial strawberry (*Fragaria* × *ananassa*) [25], Chinese cabbage (*Brassica pekinensis*) [26], potato (*Solanum tuberosum*) [27], oil palm (*Elaeis guineensis*) [28], and apple (*Malus* × *domestica*) [29].

As a common "gene-centered" method, Y1H protocols usually involve complex cDNA libraries to screen multiple TFs with a single DNA [2, 30]. Several improvements of the method have exploited the versatility of Y1H to develop more "TF-centered" approaches [31, 32]. In this chapter, we present a simplified method that is not centered on finding new prey from cDNA libraries but on testing a small-scale screening for specific target plant proteins. We also include the latest improvement of classical Y1H with Gateway cloning® and using *AUR1-C* as a reporter gene.

2 Materials

2.1 Bacterial Growth Media

Bacterial culture requires specific formulations. This protocol uses LB media plus ampicillin (LB/Amp) for cloning and plasmid DNA preparation.

1. LB/Amp plates: Weigh 10 g peptone, 5 g yeast extract, and 10 g NaCl. Add about 800 mL of deionized water and mix. Add 20 g agar powder and make up to 1 L with deionized water. Autoclave at 121 °C for 15 min. Transfer the media to a flow chamber or another sterile working area once the media has cooled to at least 50 °C. Then, add 0.50 mL of the 100 mg/mL Amp stock solution. Plate them on 100 mm Petri dishes. Store plates at 4 °C.

2.2 Yeast Growth Media

1. YPDA broth: Weigh 10 g yeast extract, 20 g peptone, and 20 g dextrose. Add about 800 mL of deionized water and mix. Add 0.16 g adenine hemisulfate and make up to 1 L with deionized water. Autoclave at 121 °C for 15 min. Store at 4 °C (*see* **Note 1**).
2. YPDA plates (solid media): Prepare 1 L of YPDA broth as in the previous solution and dissolve 20 g agar powder. Autoclave at 121 °C for 15 min and plate them on 100 mm Petri dishes. Store plates at 4 °C.
3. SD/-Leu plates: Prepare one of the premixes of yeast minimal media/synthetic-defined (SD/-Leu Broth, Cat. No. 630310), and add 10 g agar in 0.5 L deionized water. Sterilize at 121 °C for 15 min, and plate them on 100 mm Petri dishes.
4. SD/-Ura plates: Prepare one of the premixes of yeast minimal media/synthetic-defined (SD/-Ura Broth), and add 10 g agar in 0.5 L deionized water. Sterilize at 121 °C for 15 min, and plate them on 100 mm Petri dishes.

2.3 Antibiotics

1. 100 mg/mL ampicillin stock (Amp): For a stock solution, weigh 1 g of ampicillin sodium salt, and dissolve it in 10 mL of sterile deionized water. Sterilize it through a 0.22 μm syringe filter and aliquot it into 1.5 mL tubes. Store the tubes at −20 °C until ready for use.
2. 500 μg/mL aureobasidin A (AbA) stock: Dissolve 0.5 mg aureobasidin A in 1 mL absolute ethanol. Aliquot 50 μL of AbA in 0.6 mL tubes and store them at −20 °C.

2.4 Buffers

1. 1× TE/LiAc buffer: For 3 mL preparation, add 300 μL 1 M LiAc and 300 μL 10× TE buffer to 2400 μL sterile water. Use immediately.
2. 1× PEG/LiAc buffer: For 5 mL preparation, add 500 μL 1 M LiAc and 500 μL 10× TE buffer to 4000 μL PEG 50% solution. Use immediately.

2.5 Microbial Strains

1. Yeast strain: Y1HGold yeast strain. Genotype *MATα*, *ura3-52*, *his3-200*, *ade2-101*, *trp1-901*, *leu2-3,112*, *gal4Δ*, *gal80Δ*, *met–*, *MEL1* (*see* **Note 2**).
2. Bacterial strain: Use any commercially available competent *ccdB*-sensitive *Escherichia coli* strain, such as DH5α or TOP10.

2.6 Additional Materials Required

1. Plasmids: pDONR221 or pDONR/Zeo entry clone containing the transcription factor sequence of interest (*see* **Note 3**); pAbAi vector; pGADT7-GW (Plasmid #61702 Addgene) (*see* **Note 4**); p53-AbAi control vector; and p53 control insert and pGADT7-Rec (*see* **Note 5**).
2. DNA bait cloning PCR primers: Design appropriate primers to amplify your target DNA bait, and add the recognition sites for restriction cloning to the end of each primer (*see* **Note 6**).
3. Enzymes: Gateway LR clonase II enzyme; T4 DNA ligase; and the restriction enzymes *Hin*dIII, *Xho*I, and *Bbs*I or *Bst*BI (*see* **Note 7**).
4. Stock solutions: 1× TE buffer; 10% DMSO; salmon sperm DNA (10 mg/mL) (*see* **Note 8**).
5. Molecular biology reagents and kits: *Taq* polymerase, agarose, nucleic acid gel stain, DNA Ladder, 1× TAE buffer, 1× TE buffer, sterile water, gel, and PCR purification kits, like Wizard® SV Gel and PCR Clean-Up System.
6. Lab Equipment and supplies: Thermal cycler, dry bath/heat block, UV/Vis spectrophotometer, orbital shaker, constant-temperature incubator, microcentrifuge tubes, sterile flasks, 50 mL sterile centrifuge tubes, and Petri dishes.

3 Methods

3.1 DNA-Bait Cloning

The Y1H assay requires a selected target DNA sequence (DNA-bait), such as a promoter sequence, an enhancer region of a gene, or an independent *cis*-regulatory element to clone next to a reporter gene. In this method, we use a DNA-bait::*AUR1-C* construct in a pAbAi vector as a reporter of PDI (Fig. 1a). We recommend cloning the target DNA sequence into the pAbAi vector by traditional cloning using *Hin*dIII and *Xho*I restriction enzymes (*see* **Note 1**).

1. Use appropriate primers to add restriction sites for traditional cloning by PCR to the target DNA sequence. To maintain the proper orientation, we recommend adding the *Hin*dIII restriction site at the 5′ end and the *Xho*I restriction site at the 3′ end of the DNA-bait.
2. Amplify by PCR with the designed primers and purify the product by gel purification.
3. Digest at least 2 μg of the purified PCR product and 1 μg of the pAbAi vector independently using *Hin*dIII and *Xho*I restriction enzymes to obtain compatible overhangs.
4. Verify the correct size of the fragments by gel electrophoresis, and perform a gel purification of the bands.
5. Ligate both the DNA-bait and pAbAi vector with T4 DNA ligase [33] using a molar ratio of 1:3 (pAbAi: DNA-bait).
6. Transform 50 μL of competent *E. coli* cells with 3 μL of the ligation reaction.
7. Select positive colonies by plating transformed *E. coli* onto LB/Amp plates (100 μg/mL), and incubate them overnight at 37 °C.
8. Verify by sequencing that the DNA-bait is located before the *AUR1-C* gene in the pAbAi vector, in the same sense as it is placed before their controlled gene in the genome. This will be our pAbAi-DNA-bait plasmid.
9. Finally, linearize the pAbAi-DNA-bait plasmid by restriction digest with the *Bbs*I or *Bst*BI enzymes (*see* **Note 7**).

3.2 Protein-Prey Construct Cloning

1. Mix 250 ng (~85 fmol) pDONR221-TF entry clone and ~ 500 ng (~80 fmol) pGADT7-GW vector in a 0.2 mL sterile tube. Add TE buffer up to 8 μL and 2 μL Gateway LR clonase II enzyme mix. Use TE buffer instead of the pDONR221-TF entry clone as a negative control mix.
2. Vortex briefly and microcentrifuge the tube. Then, incubate for 2 h at 25 °C in a thermocycler. Overnight incubation can be used for large entry clones.

3. Use 2 μL of the mix to transform chemically competent *E. coli* cells.
4. Select positive colonies by plating transformed *E. coli* onto LB/Amp plates (100 μg/mL), and incubate them overnight at 37 °C.
5. Verify the correct insertion of the TF sequence on the pGADT7-GW vector by restriction digest and sequencing. Be aware to avoid any frameshift mutation between GAL4AD and TF.

3.3 Integrating Reporter pAbAi Constructs on the Yeast Genome

The pAbAi vectors integrate into the host genome through homologous recombination of the *ura3-52* locus of the Y1HGold yeast strain and the *URA3* gene of the pAbAi plasmid. The *URA3* gene is a selector for the yeasts successfully recombined after transformation.

3.3.1 Yeast-Competent Cell Preparation

1. Before plasmid transformation, you must make the yeast cells competent by incubating them on a buffered lithium acetate mix. Pick one colony from a recent Y1HGold yeast strain YPDA plate, and inoculate it into 50 mL of YPDA broth in a 150 mL sterile flask. Leave them to grow overnight at 30 °C with shaking at 250 rpm. Reserve some YPDA broth to use as a blank when measuring OD600.
2. The next day, check the OD600 on the stationary phase, which should be no higher than 1.5.
3. Dilute the starter culture into 200 mL YPDA broth, checking the OD600 ~0.3.
4. Grow the new culture for 3–4 h, shaking at 30 °C until it reaches OD600 ~0.6.
5. Transfer the new culture to 50 mL tubes, and harvest cells by centrifugation at room temperature for 5 min at 2600×*g*. Do not centrifuge for more time, as it will be difficult to resuspend the pellet (*see* **Note 9**).
6. Discard cleared supernatant and carefully resuspend every tube cell in 50 mL of sterile water. Then, centrifuge them at room temperature for 5 min at 2600×*g*, repeating the previous step.
7. Finally, discard the supernatant, resuspend cells in 400 μL cold 1× TE/LiAc buffer in each tube, and aliquot cells into four sterile 1.5 mL microcentrifuge tubes.
8. The cells are now ready for transformation. If not used immediately, they can be stored at −80 °C (*see* **Note 10**).

3.3.2 Plasmid Transformation

This step is carried out by transforming plasmid DNA with yeast-competent cells previously made and a carrier DNA or salmon sperm DNA using protocols previously introduced [34] with some modifications [31, 32].

1. Mix 5–10 μL salmon sperm DNA (*see* **Note 8**) previously heated at 95 °C for 5 min into a sterile 1.5 mL microcentrifuge tube and 200 ng yeast plasmid DNA (pAbAi-DNA-bait).
2. Simultaneously, for the negative PDI control, add 125 ng p53 control insert with 1 μg pGADT7-Rec linearized with *Sma*I restriction enzyme (*see* **Note 5**). Then, as a positive PDI add 200 ng yeast plasmid p53-AbAi control. Add salmon sperm DNA in both tubes, too.
3. Add 100 μL of recently made yeast-competent cells to each tube and vortex.
4. Add 600 μL 1× PEG/LiAc buffer and vortex.
5. Incubate at 30 °C for 30 min with shaking. Turn on a water bath or a heat block at 42 °C.
6. Add 70 μL of 10% DMSO. Mix by inversion four times.
7. Incubate the tubes in the 42 °C heat block for 15 min. Mix occasionally.
8. Incubate on ice for 2 min.
9. Spin for 5 s in a microcentrifuge, removing the supernatant. Resuspend cells in 200 μL sterile water.
10. Individually, plate 100 μL of transformed pAbAi-DNA-bait and the positive PDI control yeast tubes on selective SD/-Ura agar plates.
11. Plate 100 μL of transformed negative PDI control yeast tube on a selective SD/-Leu agar plate.
12. Incubate plates upside-down at 30 °C for 3 days until the first colonies appear.

3.3.3 Determine the Minimal Inhibitory Concentration (MIC) of Aureobasidin A (AbA) for the Bait Strains

In Y1H assays, AbA resistance is given by the expression of the *AUR1-C* mutant gene [15, 16]. Some bait yeast strains might grow on low AbA concentrations in the absence of positive PDI. To avoid false positives, concentrations of AbA must be titrated in increasing order according to each individual bait strain.

1. Select positive clones on the SD/-Ura plates (transformed pAbAi-DNA-bait and the positive PDI control yeast strains), picking five or six colonies, and resuspend them onto 500 μL of sterile water to finally plate 90 μL on SD/-Ura, with increasing concentrations of AbA (50, 100, 150, and 200 ng/mL). Set plates at 30 °C for 2–3 days.
2. Results must be from ~2000 colonies for SD/-Ura and a decreasing number of colonies according to AbA increasing concentrations (*see* **Note 11**).
3. Choose the minimal inhibitory concentration of AbA for PDI screening.

3.4 Y1H Screening Interaction

Screening is essential for validating protein-DNA interaction and determining whether specific proteins interact with specific promoters or any other sequence of interest, being common promoter regions or regulatory elements.

3.4.1 Transformation of Competent Bait-Reporter Yeast with Prey Construct

Selection of plasmids in yeast hybrids is based on the capacity of transformant cells to grow on a medium that lacks specific components (e.g., uracil pyrimidine or leucine amino acid) because it is complemented by a plasmid which allows its growth in a medium that lacks the component and confirms a positive transformation.

1. Repeat Subheading 3.3.1 of yeast-competent cell preparation with positive clones of Subheading 3.3.2. Pick a few colonies from the SD/-Ura plate, and inoculate them into 50 mL YPD broth in a 150 mL sterile flask. Leave them to grow at 30 °C overnight, shaking at 250 rpm and proceeding as mentioned in Subheading 3.3.1.
2. Once you have made yeast-competent cells with the pAbAi-DNA-bait and the positive PDI control yeast strains, you must carry out a plasmid transformation.
3. First, use 200 ng of the pGADT7-TF plasmid to transform 100 μL of competent pAbAi-DNA-bait yeast strain.
4. Then, add 125 ng p53 control insert with 1 μg pGADT7-Rec linearized with *Sma*I restriction enzyme (*see* **Note 7**) to 100 μL of the positive PDI control competent yeast strain.
5. Proceed with transformation on **step 4** of Subheading 3.3.2.
6. Finally, according to Table 1, you must plate 100 μL of transformed cell yeasts on selective media.
7. Incubate them at 30 °C for 3 days.

3.4.2 Screening Interaction

To confirm the positive interaction between the DNA bait sequence and the transcription factor coupled to the GAL4-activation domain in the pGADT7 vector. The AbA antibiotic resistance should be tested through increasing concentrations in SD/-Leu medium plates (Fig. 1c).

1. Select colonies of the SD/-Leu plates and dilute them onto 500 μL sterile water to finally plate six to eight small drops (5–10 μL) (*see* **Note 12**) of the resulting transformation of bait-reporter yeast with the prey construct on SD/-Ura and SD/-Leu plates.
2. Then, droplets should be plated onto plates with increasing aureobasidin A (AbA) concentrations, starting from the MIC.
3. Setting plates at 30 °C for 2–3 days (*see* **Note 13**), droplets should grow like the yeast phenotype testing (Table 1).
4. The higher the concentration, the stronger the interaction between the DNA sequence and the TF, which activates the expression of the reporter gene [2].

Table 1
Phenotype testing of yeast droplets on various YPD and SD media

Bait	Prey	YPD	YPD AbA^{200}	SD/-Leu	SD/-Ura	SD/-Leu AbA^{200}	SD/-Ura AbA^{200}
pAbAi	No prey	+	–	–	+	–	–
No bait	pGADT7	+	–	+	–	–	–
pAbAi-DNA-bait	No prey	+	–	–	+	–	–
No bait	pGADT7-TF	+	–	+	–	–	–
pAbAi-DNA-bait	pGADT7-TF[a]	+	+/–[a]	+	+	+/–[a]	+/–[a]
p53-AbAi	No prey	+	–	–	+	–	–
No bait	pGADT7-rec-p53	+	–	+	–	–	–
p53-AbAi	pGADT7-rec-p53	+	+	+	+	+	+
Empty Y1HGold yeast		+	–	–	–	–	–

[a]Phenotype depends on a positive (+) or negative (–) interaction

4 Notes

1. YPDA broth and solid media can be replaced by YPD, which lacks adenine hemisulfate; however, we do not recommend this change. YPDA is preferred to prevent an unwanted reddish coloration of the yeast colonies.
2. We recommend purchasing from Clontech. The strain is also included in the Matchmaker® Gold Yeast One-Hybrid Library Screening System kit.
3. For protein-prey cloning, the coding region of a selected transcription factor (TF) or other DNA-binding protein must first be available as a purified entry clone for Gateway cloning. We recommend using a pDONR221 or pDONR/Zeo plasmid as a donor vector for Gateway cloning because its antibiotic selection is compatible with the pGADT7-GW destination vector used in this method. Any other donor vector can be used unless it has ampicillin resistance as the selectable marker. For example, we will use a TF cloned in a pDONR221 donor vector (pDONR221-TF with kanamycin resistance) as the entry clone to generate an *ADH1*$_{pro}$*::GAL4AD-TF* construct in the pGADT7-GW destination vector (Fig. 1b).

4. The pGADT7-GW plasmid could be replaced with any alternative plasmid that combines a Gal4-activation domain with the selected TF under the control of an *ADH1* promoter or a yeast-compatible promoter. Also, a functional auxotrophic gene mutated in the Y1HGold yeast strain.
5. p53-Abai control vector and pGADT7-Rec (*Sma*I linearized) can be purchased from Clontech. The plasmids are also included in the Matchmaker® Gold Yeast One-Hybrid Library Screening System.
6. We recommend the use of restriction sites of *Hind*III and *Xho*I at 5′ end and 3′ end, which can be changed for any other within the multiple cloning site (MCS) of the pAbAi vector, respecting the directionality of the DNA-bait. Also, the DNA bait must not have an internal restriction site for the selected enzymes.
7. The DNA-bait selected to be cloned in the pAbAi vector must not have an internal restriction site for the chosen linearization enzyme (*Bbs*I or *Bst*BI).
8. This can be replaced by carrier DNA from Clontech Yeastmaker Carrier DNA denatured (10 mg/mL).
9. Centrifuge at room temperature; it will be easier to resuspend the pellet than at 4 °C.
10. It is highly recommended to use them within 2 weeks, maximum, since cells do not have glycerol.
11. If 200 ng/mL does not suppress basal expression, increasing the AbA concentration to 500–1000 ng/mL might be necessary.
12. It is strongly recommended to use 5 μL instead of 10 μL of small droplets due to the possibility of deformation due to spillage.
13. To avoid contamination, it is highly recommended that this procedure be carried out in a laminar air flow booth with negative air pressure.

Acknowledgments

This work was supported by the National Research and Development Agency (ANID, Chile) grants Millennium Science Initiative Program—NCN2021_010 and FONDECYT/Regular 1210941.

Conflict of Interest The authors have no conflicts of interest to declare that are relevant to the content of this chapter.

References

1. Sánchez-Gómez C, Posé D, Martín-Pizarro C (2022) Insights into transcription factors controlling strawberry fruit development and ripening. Front Plant Sci 13:1022369
2. Reece-Hoyes JS, Walhout AJM (2012) Gene-centered yeast one-hybrid assays. In: Methods in molecular biology. Humana Press, Totowa, pp 189–208
3. Muhammad II, Munusamy U, Kong SL et al (2020) RNA-seq and ChIP-seq as complementary approaches for comprehension of plant transcriptional regulatory mechanism. Int J Mol Sci 21:167
4. Liu Y, Du M, Deng L et al (2019) MYC2 regulates the termination of Jasmonate signaling via an autoregulatory negative feedback loop. Plant Cell 31:106–127
5. Collins GA, Lipford JR, Deshaies RJ et al (2009) Gal4 turnover and transcription activation. Nature 461:E7
6. Ezhkova E, Tansey WP (2006) Chromatin immunoprecipitation to study protein-DNA interactions in budding yeast. Methods Mol Biol 313:225–244
7. López-Vidriero I, Godoy M, Grau J et al (2021) DNA features beyond the transcription factor binding site specify target recognition by plant MYC2-related bHLH proteins. Plant Commun 2:100232
8. Franco-Zorrilla JM, Prat S (2021) DAP-Seq identification of transcription factor-binding sites in potato. In: Dobnik D, Gruden K, Ramšak Ž et al (eds) Solanum tuberosum: methods and protocols. Springer, New York, pp 123–142
9. Oliveira R, Pinho E, Sousa AL et al (2022) Improving aptamer performance with nucleic acid mimics: de novo and post-SELEX approaches. Trends Biotechnol 40:549–563
10. Song L, Koga Y, Ecker JR (2016) Profiling of transcription factor binding events by chromatin immunoprecipitation sequencing (ChIP-seq). Curr Protoc Plant Biol 1:293–306
11. Kaufmann K, Muiño JM, Østerås M et al (2010) Chromatin immunoprecipitation (ChIP) of plant transcription factors followed by sequencing (ChIP-SEQ) or hybridization to whole genome arrays (ChIP-CHIP). Nat Protoc 5:457–472
12. Kansal S, Panwar V, Mutum RD et al (2021) Investigations on regulation of MicroRNAs in rice reveal [Ca2+]cyt signal transduction regulated MicroRNAs. Front Plant Sci 12:720009
13. Godoy M, Franco-Zorrilla JM, Pérez-Pérez J et al (2011) Improved protein-binding microarrays for the identification of DNA-binding specificities of transcription factors. Plant J 66:700–711
14. Reece-Hoyes JS, Marian Walhout AJ (2012) Yeast one-hybrid assays: a historical and technical perspective. Methods 57:441–447
15. Kastuhisa U, Tamio H, Hideyo Y et al (1993) Biological properties of Aureobasidin A, a cyclic depsipeptide antifungal antibiotic. J Antibiot (Tokyo) 46:1414–1420
16. Hashida-Okado T, Ogawa A, Endo M et al (1996) AUR1, a novel gene conferring aureobasidin resistance on Saccharomyces cerevisiae: a study of defective morphologies in Aur1p-depleted cells. Mol Gen Genet 251:236–244
17. Hashida-Okado T, Ogawa A, Kato I et al (1998) Transformation system for prototrophic industrial yeasts using the AUR1 gene as a dominant selection marker. FEBS Lett 425: 117–122
18. Sewell JA, Fuxman Bass JI (2018) Options and considerations when using a yeast one-hybrid system. In: Oñate-Sánchez L (ed) Two-hybrid systems: methods and protocols. Springer, New York, pp 119–130
19. Park S, Xu Y, Stowell XF et al (2006) Limitations of yeast surface display in engineering proteins of high thermostability. PEDS 19: 211–217
20. Nightingale DJ, Geladaki A, Breckels LM et al (2019) The subcellular organisation of Saccharomyces cerevisiae. Curr Opin Chem Biol 48: 86–95
21. Kim H, Yoo SJ, Kang HA (2015) Yeast synthetic biology for the production of recombinant therapeutic proteins. FEMS Yeast Res 15: 1–16
22. Stark MJR (1998) 14 studying posttranslational modifications in yeast: intracellular proteins. In: Brown AJP, Tuite M (eds) Methods in microbiology. Academic Press, London, pp 239–254
23. Smale ST (2010) Luciferase Assay. Cold Spring Harb Protoc 2010:pdb.prot5421
24. Yu X, Wang L, Xie Y et al (2024) OsBBP1, a newly identified protein containing DUF630 and DUF632 domains confers drought tolerance in rice. Plant Sci 345:112119
25. Valenzuela-Riffo F, Delgado C, Morales-Quintana L et al (2024) The strawberry transcription factor FaWRKY48 transactivates the FaNCED1 promoter as revealed by yeast-one hybrid and Nicotiana benthamiana transfection assays. Sci Hortic 323:112545

26. Ge W, Zhang J, Feng H et al (2023) The root meristem growth factor BrRGF6 positively regulates Chinese cabbage to infection of clubroot disease caused by Plasmodiophora Brassicae. Hortic Res 10:uhac292
27. Han Y, Yang R, Zhang X et al (2024) MYB24, MYB144, and MYB168 positively regulate suberin biosynthesis at potato tuber wounds during healing. Plant J 119(3):1239–1257
28. Luo S, Huang J, Jin L et al (2024) Transcription factor EgGRP2A regulates EgFATA expression and promotes oleic acid accumulation in oil palm (Elaeis guineensis). J Plant Physiol 299:154263
29. Zheng L, Ma W, Liu P et al (2024) Transcriptional factor MdESE3 controls fruit acidity by activating genes regulating malic acid content in apple. Plant Physiol. https://doi.org/10.1093/plphys/kiae282
30. Reece-Hoyes JS, Diallo A, Lajoie B et al (2011) Enhanced yeast one-hybrid assays for high-throughput gene-centered regulatory network mapping. Nat Methods 8:1059–1064
31. Ji X, Wang L, Zang D et al (2018) Transcription factor-centered yeast one-hybrid assay. In: Methods in molecular biology. Humana Press, New York, pp 183–194
32. Jingwen W, Jingxin W, Ye Z et al (2023) Building an improved transcription factor-centered yeast one hybrid system to identify DNA motifs bound by protein comprehensively. BMC Plant Biol 23:1–11
33. Gaastra W, Hansen K (1985) Ligation of DNA with t(4) DNA ligase. Methods Mol Biol 2:225–230
34. Ito H, Fukuda Y, Murata K et al (1983) Transformation of intact yeast cells treated with alkali cations. J Bacteriol 153:163–168

Chapter 7

Targeted DNA Affinity Purification for Quantifying DNA-Binding Specificities of Transcription Factors

Eva Álvarez and José M. Franco-Zorrilla

Abstract

Plant growth relies on flexible gene regulation to adapt to environmental changes. This process is ultimately controlled by transcription factors (TFs), which bind to specific DNA motifs, known as TF-binding sites (TFBS), located in the gene regulatory regions to regulate their expression. These interactions play crucial roles in plant development and responses to environmental cues, as well as in plant evolution and domestication, making both *cis*- (i.e., TFBS) and *trans*-regulatory factors as potential molecular targets in plant breeding for traits such as yield, quality, and stress resilience. These biotechnological approaches require precise knowledge of the target gene sets and TFBS specifically recognized by TFs. Recent advances in high-throughput sequencing techniques have enabled precise identification of TF target genes, especially thanks to methodologies that combine the main features of both in vitro and in vivo approaches. However, small scale and targeted approaches are still needed to evaluate the relative contribution of specific nucleotide positions in TF recognition. In this chapter, we describe a modified version of DNA Affinity Purification sequencing (DAP-seq) that replaces genomic DNA with a PCR-generated library of TFBS variants. This approach, termed targeted-DAP, allows a flexible and quantitative analysis of TF-binding using next-generation sequencing. Additionally, expressing TFs in *Escherichia coli* provides an economical source of proteins, enabling scalable and cost-effective analysis of DNA-binding specificities. We showed the benefits of this technique to demonstrate the contribution of the genomic context around the TFBS for specific recognition of a bHLH TF. Development of targeted DAP-seq would be of interest for the evaluation of nucleotide variation—either allelic or generated by CRISPR/Cas—within TFBS in TF recognition with predictable consequences on plant phenotypes.

Key words Transcription factor, TF-binding site (TFBS), DNA-binding specificity, DNA affinity purification (DAP-seq), Bound amplicon, Mutant amplicon

1 Introduction

As sessile organisms, plants need to adapt to environmental conditions in order to grow and develop. Such plasticity is possible, thanks to the existence of signaling pathways that allow the regulation of gene expression levels. Genes are activated or repressed through the action of regulatory proteins, the transcription factors (TFs). These bind directly to TF-binding sites (TFBS), short DNA

José M. Franco-Zorrilla and Carmen Martín-Pizarro (eds.), *Plant Transcription Factors: Methods and Protocols*, Methods in Molecular Biology, vol. 2985, https://doi.org/10.1007/978-1-0716-4972-5_7,

motifs (6–12 bp long), located in the regulatory regions of the target genes, commonly upstream of the transcription start site (TSS). The importance of TFBS in specifying cell functions is well known, and diversification in these motifs has been linked to phenotypic diversity and crop evolution. Additionally, specification of desired phenotypical traits such as higher yield, fruit quality, or improved environmental adaptation is proven to be dependent on several TFs. This highlights the relevance of variation of both *cis*- (i.e., TFBS) and *trans*- (TF) regulators during crop domestication [1, 2].

In the light of these findings, the biotechnological potential of TFs becomes evident. Given their hierarchical mode of action, whole signaling pathways can be modified through genetic manipulation of just one—or a few—loci. This would expand the possibilities of generating new plant varieties with desirable agronomical traits. Genetic manipulation is not limited to TF-coding genes (i.e., *trans*-regulation) but also to promoters or TFBS (*cis*-regulation). Together, these examples demonstrate that phenotypic diversity during evolution and plant breeding is driven by both *cis*- and *trans*-regulators. Therefore, their study is relevant to uncover the genetic basis of plant adaptation to changing natural environments.

During the last years, the study of transcriptional regulation has experienced a significant advance, thanks to an increase in sample processing and affordability of high-throughput technologies, leading to the identification of target genes for many animal and plant TFs. Among others, DNA Affinity Purification and sequencing (DAP-seq) [3, 4] is a precise and relatively fast methodology to find target genes for a given TF. This technique is based on the purification of DNA-protein complexes that are formed in vitro, followed by the elution of the bound DNA, which is then sequenced. As it uses plant genomic DNA (gDNA), it combines the advantages of both in vitro and in vivo approaches [3, 4].

In order to evaluate to what extent the relative binding affinities of TFs to their corresponding TFBS imply a change in their target genes, here we perform a modification of the DAP-seq method that we termed targeted-DAP. Instead of gDNA, we use a PCR-amplified mix of fragments containing several variants of TFBS. This approach follows the same principle as the Electrophoretic Mobility-Shift Assays (EMSA) since this amplicon library would act the labeled probes to which the TF is bound in EMSA. Although EMSA does not rely on specialized equipment and is easy to carry out, the number of probes that can be tested at once is limited, and it only provides qualitative information. This targeted-DAP approach performed on a pool of amplicons allows scanning tens of different probes in a single experiment, reducing experimental costs. Additionally, next-generation sequencing (NGS) enables the quantification of the relative TF-binding affinities, as the number of reads associated with the binding to each

TFBS can be easily quantified. Furthermore, in this protocol, we also propose the expression of the TF of interest in *Escherichia coli* and the use of whole protein extracts as a source for recombinant proteins, instead of transcription/translation coupled systems. This modification allows performing several experiments in parallel in a cost-effective manner, as it has been applied previously for the characterization of DNA-binding specificities [5–8].

2 Materials

2.1 Reagents and Solutions

1. Elution buffer (EB): 10 mM Tris–HCl pH 8, 1 mM EDTA pH 8.
2. Ethanol.
3. 100 mM dNTP PCR Grade Set.
4. 1% NP-40.
5. 10× Phosphate-buffered saline (PBS): 1.37 M NaCl, 27 mM KCl, 100 mM Na_2HPO_4, 18 mM KH_2PO_4.
6. 0.1 M Phenylmethanesulfonyl fluoride (PMSF) in ethanol.
7. LB medium supplemented with 0.2% glucose.
8. 1 M Isopropyl β-D-thiogalactoside (IPTG).
9. 10× Annealing buffer: 100 mM Tris, pH 8.0, 500 mM NaCl, 10 mM EDTA.
10. Agarose and TBE 10× (0.9 M Tris, 0.9 M H_3BO_3, 0.02 M NaEDTA) for agarose gel electrophoresis.
11. 30% Acrylamide/Bis (29:1), 1.5 M Tris-HCl pH 8.8, 10% SDS, 10% ammonium persulfate (APS), N,N,N′,N′--Tetramethyl ethylenediamine (TEMED) for polyacrylamide-SDS gel electrophoresis (SDS-PAGE).

2.2 Enzymes and Kits

1. Phusion High-Fidelity DNA Polymerase (NEB).
2. Silica membrane-based columns for DNA purification from agarose gels.
3. Klenow Polymerase (3′ → 5′ exo-) (NEB).
4. T4 DNA Ligase (Promega).
5. Amylose magnetic beads 25 mg, 10 mg/mL (NEB).
6. Qubit™ 1× dsDNA HS Assay Kit (Thermo Fisher).
7. Agencourt AMPure XP beads (Beckman Coulter).

2.3 Oligonucleotides

1. Adaptor Strand A: 5′— ACACTCTTTCCCTACACGACGCTCTTCCGATC *T-3′ (asterisk indicates phosphorothioate bond) (*see* **Note 1**).

2. Adaptor Strand B: 5′— P- GATCGGAAGAGCACACGTCT GAACTCCAGTCAC-3′ (“P” indicates a 5′ phosphate group) (*see* **Note 1**).
3. 30 μM Y-adaptor (*see* **Note 2**).
4. Dual Index Multiplex Primers for Illumina: i5, 5'- AATGA TACGGCGACCACCGAGATCTACAC NNNNNNNN ACAC TCTTTCCCTACACGACGCTCTTCCGATC *T – 3′; i7, 5′- CAAGCAGAAGACGGCATACGAGAT NNNNNNNN GTGA CTGGAGTTCAGACGTGTGCTCTTCCGATC *T -3′. NNNNNNNN” represents the 8mer index sequence for multiplexing. Asterisk indicates a phosphorothioate bond (*see* **Note 3**).

2.4 Equipment

1. Qubit 4 Fluorometer (Thermo Fisher).
2. Nanodrop spectrophotometer (Thermo Fisher).
3. Tabletop microcentrifuge.
4. Thermocycler.
5. Agarose gel electrophoresis device.
6. Polyacrylamide gel electrophoresis (PAGE) device.
7. Neodymium magnet or magnetic rack suitable for 1.5 mL tubes.
8. Tube rotator mixer.
9. Orbital platform shaker.
10. Probe sonicator for cell disruption.
11. Incubator shaker.

3 Methods

3.1 Amplicon Pool Generation

3.1.1 Design of Amplicons

This protocol starts with the preparation of the DNA amplicons spanning the expected sequence motif recognized by the TF of interest, as well as their corresponding mutant versions. Point mutations in TFBS are incorporated in a two-step PCR protocol, using complementary primers carrying the desired nucleotide modifications and plant gDNA as template.

1. Recommended amplicon size in the range of 200 bp, with target sequence motif in the center of the amplicon. Outer primers can be designed using any of the available primer design tools, such as NCBI's Primer-BLAST (https://www.ncbi.nlm.nih.gov/tools/primer-blast/).
2. Design two partially overlapping oligonucleotides including the point mutation(s) to incorporate in the mutant amplicon (Fig. 1) (*see* **Note 4**).

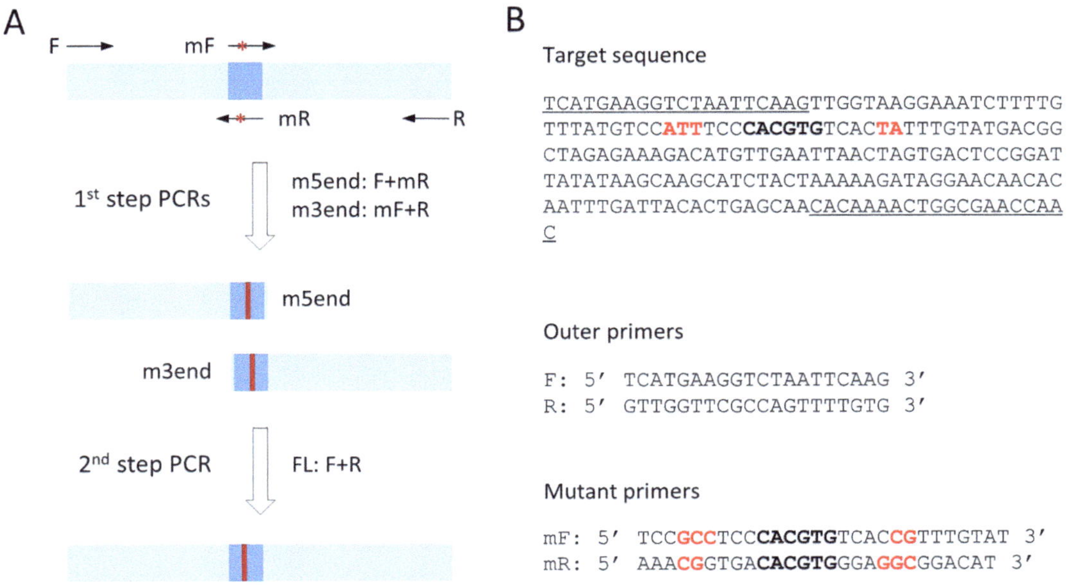

Fig. 1 Generation of mutant amplicons. (**a**) Design of a two-step PCR to introduce single-point mutations (represented as red asterisks) in the amplicon sequence. (**b**) Example of an amplicon that includes a TF-binding sequence in bold (G-box, 5′-CACGTG-3′) and the nucleotide positions to modify in the mutant version of the amplicon in red. Outer and mutant primers used in the two-step PCR are indicated

3.1.2 Standard and Two-Step PCR Reactions

1. Assemble PCR reactions to generate wild-type amplicons using outer F and R primers and gDNA as template (Fig. 1, Tables 1 and 2). Keep PCR products for further verification.
2. To obtain the amplicons containing point mutations, perform two independent PCR reactions with the primer combinations F + mR and mF + R (Fig. 1; Table 1) to generate m5end and m3end fragments, respectively. Set up standard PCR reactions as in **step 1**, but for 25 cycles (Table 2).
3. Separate m5end and m3end PCR products in 1.5% agarose gel electrophoresis, cut agarose slices, and purify through silica membrane-based columns. Quantify eluted DNA with Nanodrop.
4. Assemble the second-step PCR reaction with the same elements and cycling conditions as in **step 1**, but using as template 25 ng of each m5end and m3end fragments and outer F and R primers (Tables 1 and 2).
5. Separate PCR products from **steps 1** and **4** in agarose gel electrophoresis. Cut agarose slices for amplicons with expected size and purify through silica membrane-based columns.
6. Quantify purified PCR products with Qubit 4 fluorometer, and prepare the amplicon pool (*see* **Note 5**) by mixing all the PCR products to be used in the binding assays, including wild-type and mutant versions from several target promoters.

Table 1
DNA amplification reactions for wild-type and mutant amplicons

	Wild-type amplicon	m5end fragment	m3end fragment	Mutant amplicon
gDNA	50 ng	50 ng	50 ng	–
DNA m5end fragment	–	–	–	25 ng
DNA m3end fragment	–	–	–	25 ng
5× Phusion buffer	10 μL	10 μL	10 μL	10 μL
10 mM dNTPs	1.0 μL	1.0 μL	1.0 μL	1.0 μL
Primer F 25 μM	1.0 μL	1.0 μL	–	1.0 μL
Primer R 25 μM	1.0 μL	–	1.0 μL	1.0 μL
Primer mR 25 μM	–	1.0 μL	–	–
Primer mF 25 μM	–	–	1.0 μL	–
Phusion DNA polymerase (2 U/μL)	0.5 μL	0.5 μL	0.5 μL	0.5 μL
ddH_2O	Up to 50 μL	Up to 50 μL	Up to 50 μL	Up to 50 μL

Table 2
Thermal cycling conditions for wild-type and mutant amplicons

	Wild-type amplicon	m5end fragment	m3end fragment	Mutant amplicon
95 °C, 2 min	1 cycle	1 cycle	1 cycle	1 cycle
95 °C, 15 s; 55 °C, 30 s; 72 °C, 30 s	35 cycles	25 cycles	25 cycles	25 cycles
72 °C, 5 min	1 cycle	1 cycle	1 cycle	1 cycle
4 °C	HOLD	HOLD	HOLD	HOLD

3.2 Preparation of an Amplicon DNA Library (See Note 1)

In this step, a library of the amplicon pool compatible with the sequencing platform is prepared. By difference to the standard protocols for preparation of DNA libraries, here we omit the DNA-end repair step since the Phusion DNA Polymerase generates blunt-ended products. In total, 5 micrograms of the pool in a final volume of 32 μL are needed to start preparing the library by following this protocol. For larger volumes, the mix can be concentrated in a speed-vac.

1. Perform the A-tailing in a 50 μL reaction, by adding the following reagents to the 32 μL of the amplicon pool: 5 μL

10× NEBuffer2, 10 μL 1 mM dATP, and 3 μL Klenow fragment (3′ → 5′ exo-; 5 U/μL).

2. Incubate the reaction at 37 °C for 30 min.
3. Clean up with 50 μL (1.0×) AMPure XP Beads (*see* **Note 6**). Elute DNA in 30 μL of EB.
4. Assemble the adaptor ligation reaction by adding the following reagents to the 30 μL of eluted DNA: 5 μL T4 DNA Ligase 10× Buffer, 5 μL T4 DNA Ligase (3 U/μL), and 10 μL 30 μM annealed Y adaptor (*see* **Note 2**).
5. Incubate overnight at 16 °C.
6. Clean up with 50 μL (1.0×) AMPure XP Beads (*see* **Note 6**). Elute DNA in 31 μL of EB.
7. Use 1 μL for quantitation in a Qubit 4 fluorometer, using Qubit 1× dsDNA HS Assay Kit. The final yield of the amplicon library should be in the 2.0–2.5 μg range.

3.3 Preparation of Input DNA Sample

This step is required after the generation of every DNA library (**step** 7, Subheading 3.2) to verify the adaptor ligation through direct amplification of the amplicon pool. In addition, the amplified input sample is sequenced along with the targeted-DAP samples to be used as a negative (non-enriched) control for the affinity-purified samples. Since a single input sample serves for all the proteins/experiments that share the same library, this step is not necessary for every binding assay.

1. A PCR reaction is assembled by mixing 25 ng of the DNA library, 10 μL of 5× Phusion HF Buffer, 2.5 μL of 10 mM dNTPs, 1 μL of 25 μM i5 dual Index Multiplex Primer, 1 μL of 25 μM i7 dual Index Multiplex Primer, 1 μL of 2 U/μL Phusion DNA Polymerase, and sterile water up to 50 μL.
2. The tube is placed in a thermocycler and the following program is carried out: 95 °C for 2 min, 98 °C for 30 s, ten cycles of 98 °C for 15 s, 60 °C for 30 s, and 72 °C for 2 min, followed by final extension at 72 °C for 10 min.
3. Clean up with 50 μL (1.0×) AMPure XP Beads (*see* **Note 6**). Elute DNA in 31 μL of EB.
4. Use 1 μL for quantitation in a Qubit 4 fluorometer. Typical input sample may yield 20–30 ng/μL.

3.4 Recombinant Protein Extracts

DNA-binding assays can be performed with fusion proteins of the TF of interest to HaloTag affinity tag sequence, generated in vitro in a transcription/translation coupled system, as described [3, 4]. Alternatively, GST-tagged fusion proteins expressed in *E. coli* may also be used [3]. In this protocol, we recommend the use of N-terminal fusions to Maltose Binding Protein (MBP), since it improves the solubility and promotes the proper folding of its fusion partners [9]. This system has been particularly suitable when *E. coli* lysates are used for protein binding to DNA.

1. Induce the expression of TF-MBP fusion by the addition of IPTG to a final concentration of 1 mM in a 30 mL *E. coli* culture at 16 °C overnight (*see* **Note** 7).
2. Run a small scale non-IPTG-induced culture in parallel.
3. Save 1 mL aliquots of induced and non-induced cultures. Pellet cells by centrifugation at top speed for 1 min and discard supernatants. Store aliquots at −20 °C until use (*see* **Note 8**).
4. Harvest cells by centrifugation at 3500 × g for 20 min at 4 °C in 50 mL conical bottom tubes, and discard the supernatant.
5. Snap freeze the pellet and store at −80 °C, if it is not being used immediately.
6. The day of use, resuspend the frozen pellet by vortexing in 1 mL 1× PBS supplemented with 1 mM PMSF. Transfer the lysate to a 2 mL centrifuge tube and keep on ice.
7. Disrupt cells with a probe sonicator three times for 30 s each, keeping the tube on ice. Let stand the tube on ice for 1 min between every round of sonication to avoid overheating of the sample.
8. Centrifuge the lysate at 15,000 × g for 20 min at 4 °C and transfer the supernatant to a new tube. Repeat this step once again.
9. Keep the cleared extract on ice and use immediately for **step 1** in Section 3.5.2.

3.5 Affinity Purification of Protein-Bound DNA

The detailed procedure is an adaptation of the "Small-scale DAP-seq protocol using *E. coli*-expressed recombinant protein" by [3, 4], to be used with *E. coli* extracts expressing MBP-fusion proteins [8, 10]. Here, we recommend performing several protein-DNA-binding assays in parallel to test different numbers of amplification cycles after the binding step (*see* **Note 9**). The protocol detailed below corresponds to a single protein-DNA-binding assay.

3.5.1 Preparing Amylose Magnetic Beads

1. Transfer 25 μL of magnetic beads into a 1.5 mL microcentrifuge tube per protein extract.
2. Set the tube in a magnet for 1 min and remove the supernatant.
3. Wash the beads with 500 μL of 1× PBS and put it back to the magnet and remove the supernatant.
4. Repeat the wash step two more times. After the last wash, remove as much supernatant as possible.

3.5.2 Binding of Amylose-Fusion Protein

1. Out of the magnet, add 200 μL of the protein extract to the washed beads and resuspend completely by gentle pipetting.
2. Place the tube on a rotator mixer for 1 h 30 min at 4 °C to facilitate the protein binding to the beads.
3. Briefly spin down by centrifugation at low speed (2 s at 2000 × g), place on the magnet, and remove the supernatant.
4. Wash the beads four times in total by adding 500 μL 1× PBS + NP-40 (0.005%). Place the tubes on the magnet and remove the supernatant between washes. After the last wash, remove as much supernatant as possible.
5. Wash the beads twice with 500 μL 1× PBS without detergent, removing as much supernatant as possible after the last wash.

3.5.3 Binding of DNA to the Protein

1. Add 40 μL of 1× PBS and gently resuspend with the beads.
2. Dilute 100 ng of the amplicon library in 40 μL 1× PBS and add to the beads.
3. Incubate for 1 h at room temperature with continuous shaking. To keep the beads in suspension and allow protein-DNA binding, shake the tubes horizontally.
4. Wash the beads three times in total by adding 200 μL 1× PBS + NP-40 (0.005%). Place the tubes on the magnet and remove the supernatant between washes. Remove as much supernatant as possible after the last wash.
5. Perform two additional washes with 200 μL 1× PBS without detergent. During the second wash, transfer the beads in solution to a new 1.5 mL microcentrifuge tube.
6. Place the tubes on the magnet and discard as much supernatant as possible.
7. Add 25 μL of EB to the beads and resuspend thoroughly by vortexing.
8. Incubate the tubes in a thermal block at 98 °C for 10 min.
9. Briefly spin down the tubes, place on the magnet, and recover the supernatant containing the protein-bound DNA.

3.5.4 Amplification of Protein-Bound DNA

This step will amplify the DNA bound to the protein of interest and will attach the indexed sequencing primers. Therefore, it is critical to use unique index combinations for each sample to allow pooling several samples for sequencing, including Input library sample (Subheading 3.3). Be sure to record the index sequences used, as these will be the sample identifiers after demultiplexing.

1. Assemble PCR reaction by mixing 25 μL of enriched DNA, 10 μL of 5× Phusion HF Buffer, 2.5 μL of 10 mM dNTPs, 1 μL of 25 μM i5 Dual Index Multiplex Primer, 1 μL of 25 μM i7

Dual Index Multiplex Primer, 1 μL of 2 U/μL Phusion DNA Polymerase, and 9.5 μL of sterile water.

2. Place the tubes in a thermocycler and run the following program: 95 °C for 2 min, 98 °C for 30 s, N cycles of 98 °C for 15 s, 60 °C for 30 s, and 72 °C for 2 min, followed by extension at 72 °C for 10 min. N represents the number of cycles of amplification (*see* **Note 9**).
3. Clean up with 50 μL (1.0×) AMPure XP Beads (*see* **Note 6**).
4. Elute DNA in 31 μL of EB and use 1 μL for quantitation in a Qubit 4 fluorometer. Typical libraries may yield 20–30 ng/μL.
5. Prepare the pool of libraries to the concentration required for the sequencing provided. Be sure to mix identical quantities of the sublibraries to maintain an even representation for all of them.

3.6 DNA Sequencing and Data Analysis

1. Sequence the pool of libraries by next-generation sequencing on the Illumina platform. The read configuration will depend on the average size of the amplicons. We recommend sequencing 150 nucleotides in case of analyzing amplicons in the range of 200 bp, either in a single-read or in paired-end configuration. The read number depends on the desired coverage and the total number of libraries and the different amplicons contained in each library (*see* **Note 10**).
2. Sequence data in FASTQ format conveniently demultiplexed may be directly analyzed without the need of trimming adaptor sequences and mapping to a reference genome. Simple bash scripts can be executed in a terminal to quantify the number of reads corresponding to each of the amplicons.
3. As an example, a simple script to quantify the number of reads corresponding to the wild-type and mutant amplicons containing a G-box is provided (amplicons Gb-1^{wt} and Gb-1^{mut} in Fig. 2). Open a terminal in your computer and type:

```
cat my_bound_library1_file.fastq | grep -E 'CCATTTCCCACGTGT-
CACTATT| AATAGTGACACGTGGGAAATGG' | wc -l
cat my_bound_library1_file.fastq | grep -E 'CCGCCTCCCACGTGT-
CACCGTT| AACGGTGACACGTGGGAGGCGG' | wc -l
```

These commands will summarize the read counts in the "`my_bound_library1_file.fastq`" file containing the central sequence that differs between the wild-type and mutant Gb-1 amplicons. Given that the libraries are not directional, reverse and complementary sequences need to be queried to fully recover the reads corresponding to each amplicon.

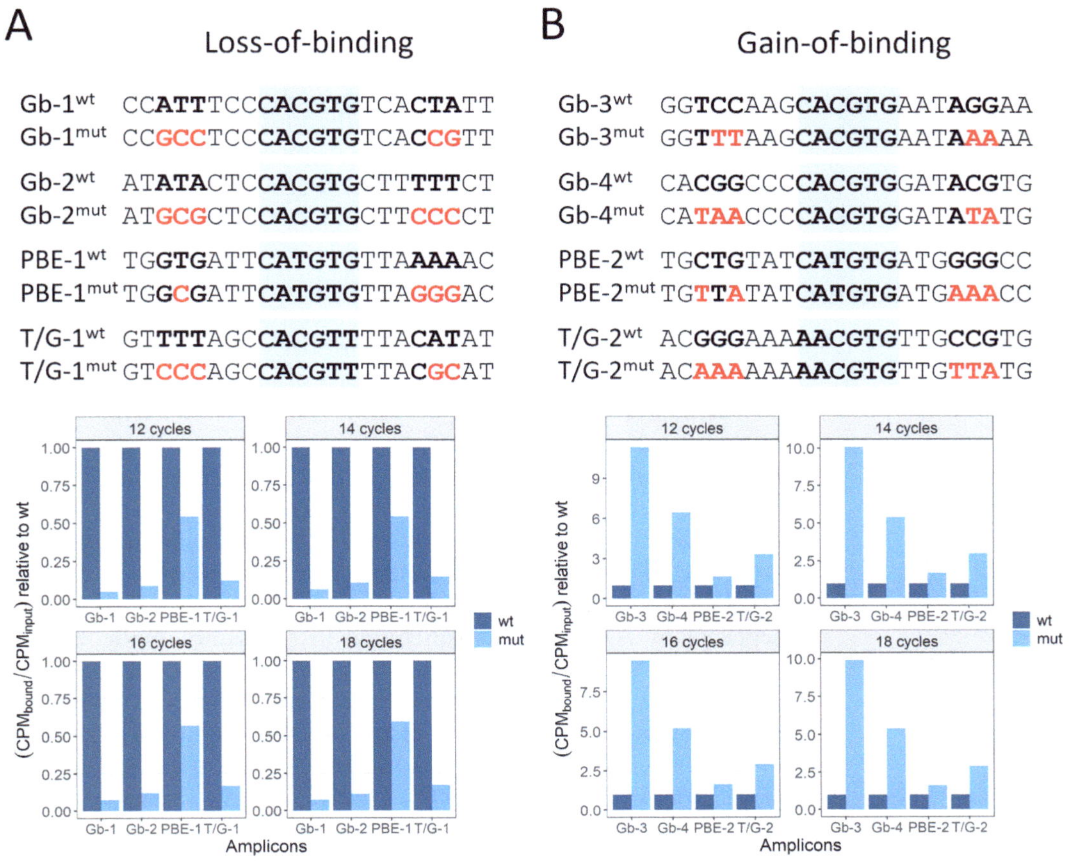

Fig. 2 (**a**) Loss-of-binding assays of MYC3 by targeted-DAP analysis. Top, amplicons containing the MYC3-binding G- (5′-CACGTG-3′), PBE (5′-CATGTG-3′) and T/G-boxes (5′-CACGTT-3′) and their corresponding mutant versions where A/T nucleotides at positions ±4, 5, and 6 were replaced by G/C. Bottom, relative binding of MYC3 to mutant amplicons in targeted-DAP assays after several cycling conditions of DNA amplification. (**b**) Gain-of-binding assays of MYC3 by targeted-DAP analysis. Same as in (**a**) but G/C nucleotides at positions were replaced by A/T. (Adapted from [10])

4. Repeat this step for every different amplicon by replacing the sequences by the central motifs characteristic for each amplicon.
5. Repeat the steps above for the "input" library to quantify the representation of the amplicons in the pool.
6. Transform raw counts to counts per million (CPM), and calculate binding efficiency to each amplicon as a ratio `CPMbound/CPMinput`.

3.7 Expected Results

The aim of this assay is to quantify the relative differences in binding affinity of a transcription factor of interest to a set of slightly different sequences. Biologically, this assay provides an experimental framework to assess the relative contribution of specific nucleotide positions within a DNA sequence for recognition of a transcription factor. In the example shown here, we assessed the relative contribution of short A/T-rich modules located outside the G-box for specific binding of the bHLH protein MYC3, involved in the initial steps of jasmonate signaling [10, 11]. These A/T-rich stretches were initially identified as important motifs for protein-DNA interaction, given their preponderance in the MYC3-bound sites, their influence in the DNA-shape, and their phylogenetic conservation [10]. Therefore, we generated a pool of amplicons containing several MYC3-binding motifs (either G-boxes, PBE, and T/G-boxes), along with their corresponding point mutations, in which A/T nucleotides at positions ±4, 5, or 6 were replaced by G/C (Fig. 2a). Targeted-DAP revealed that these point mutations in the G- and T/G-containing amplicons reduced binding between 20- and eight-fold relative to their corresponding wild-type versions and 1.8-fold in the case of the PBE-containing DNA (Fig. 2a). Following the same rationale, we generated a set of amplicons in which G/C nucleotides at positions ±4, 5, or 6 were replaced by A/T and were included in the same amplicon pool, along with their corresponding wild-type versions (Fig. 2b). By these means, we observed between 3.3- and 11.3-fold increase in MYC3 biding to the mutant versions to the amplicons containing G- and T/G-boxes and 1.7-fold in the PBE-containing amplicon (Fig. 2b). We did not observe large differences in relative binding among the different amplification conditions, but, in general, the lower the number cycles, the largest the binding differences between wild-type and mutant amplicons, supporting the benefits of performing several targeted-DAP assays in parallel covering different dynamic ranges (Fig. 2).

3.8 Biological Potential

Identification of TFBS represents a hallmark in the study of agriculturally relevant traits, as plant domestication and crop improvement largely occurred on *cis*-regulatory regions for candidate genes [2, 12]. Therefore, TFBS and *cis*-regulatory regions provide new targets with important biotechnological and agricultural potentials, given that mutations in these regions generally have weaker phenotypic effects and fewer pleiotropic consequences than in protein-coding regions, which makes them more desirable in agriculture [13, 14]. This potential has been evidenced recently by generating novel mutations that do not exist in natural populations with an impact on agricultural traits in tomato and maize [15–17]. Recent advances on CRISPR/Cas tools provide the basis for smart engineering of TFBS to generate allelic series with novel desirable phenotypes [18]. Previous knowledge of the expected impact of nucleotide variation within TFBS in TF recognition could tighten the number of genome-edited plants for phenotypic evaluation. With this regard, rapid quantification of the impact of single or multiple point mutations in TFBS located in *cis*-regulatory regions on TF-binding, such as provided by targeted-DAP technique, would accelerate the generation of novel and improved crop varieties.

4 Notes

1. Recommended high-performance liquid chromatography (HPLC) purification.
2. To prepare a 30 μM dilution of Y-adaptor, Adaptors Strand A and Strand B are dissolved with sterile water to 100 μM. Then, a mixture containing 75 μL of each diluted oligonucleotide, 25 μL of 10× annealing buffer, and 75 μL of sterile water to a final volume of 250 μL is prepared. This mixture is incubated in a thermal block for 5 min at 95 °C, and then the heat is switched off and allowed to cool down to room temperature (should take 60–90 min).
3. For reference, sequences corresponding to index i5/i7 1–6 are as follows:

i501	TATAGCCT
i502	ATAGAGGC
i503	CCTATCCT
i504	GGCTCTGA
i505	AGGCGAAG
i506	TAATCTTA
i701	CGAGTAAT
i702	TCTCCGGA
i703	AATGAGCG
i704	GGAATCTC
i705	TTCTGAAT
i706	ACGAATTC

4. An overlap of 18–20 bases between the two primers is recommended. Point mutations in the primers should be located beyond the fifth to sixth position of the oligonucleotides starting at the 3′ end, to avoid interfering in DNA amplification (Fig. 1).
5. To obtain the amplicon pool, quantities for each fragment are adjusted in order to achieve an even representation for all of them. It is important to note that UV absorbance systems (such

as Nanodrop), although simple and quite precise for DNA quantification, are also prone to quantify contaminants and salt excess in solution. Whenever possible, Qubit fluorometric quantitation using the specific assay for dsDNA is recommended.

6. Purification with AMPure XP Beads is performed following manufacturer instructions, as follows:
   - Mix DNA and beads by pipetting five to ten times and incubate for 15 min at room temperature.
   - Place the tube in a magnetic rack for at least 2 min until the solution becomes clear, and carefully remove supernatant.
   - Out of the magnet, wash the beads with 500 μL of 80% freshly prepared ethanol. Resuspend the beads by gently pipetting.
   - Place the tube in the magnet, let it stand for 1 min, and remove the supernatant.
   - Repeat the wash step with 80% ethanol, and after removing the supernatant, leave the beads at room temperature for 5 min to completely dry.
   - Resuspend the beads by pipetting in 32 μL of EB and then incubate for 10 min.
   - Place the tube in the magnet and transfer the supernatant containing the DNA to a new tube.
7. It is recommended to stick to standard methodologies for protein expression in *E. coli*. In the case of MBP-fused proteins, follow pMAL Protein Fusion and Purification System instructions (New England Biolabs). Expression levels and solubility may differ between different recombinant proteins. In these cases, it is recommended to check different inductive conditions (IPTG concentration, growing temperature, and time) to optimize protein expression. In our experience, conditions indicated in this protocol work well for most proteins.
8. Expression and solubility of recombinant proteins in the final culture must be checked by 8–12% SDS-PAGE before performing the DNA-binding assays. As a guide, please follow the steps below:
   - Resuspend frozen aliquots from **step 3** by vortexing in 0.2 mL 1×PBS supplemented with 1 mM PMSF.
   - Disrupt cells with a probe sonicator for 30 sec each, keeping the tube on ice.
   - Centrifuge at 15,000 × g for 15 min at 4 °C. Transfer the supernatant to a new centrifuge tube.

- Resuspend pellets by vortexing with 0.2 mL 1×PBS, and add to all the samples the corresponding volume of SDS-PAGE sample buffer.
- Heat samples at 95 °C for 10 min and load to gels. The recombinant protein should be detected after Coomassie staining of soluble protein extracts.

9. Typical DAP-Seq assays involve the amplification of the protein-bound DNA for 20 cycles. In this targeted-DAP protocol, we recommend a lower number of cycles for more accurate quantification of TF-bound DNA, given that the amplicon pool is not as complex as a gDNA library. We suggest performing several reactions in parallel with increasing number of cycles from 10 to 20. Nevertheless, we have not observed large differences in relative TF-DNA binding in this range of amplification cycles.
10. Several libraries indexed with unique i5/i7 primers can be pooled in a single sequencing project. Regarding the coverage of each library, this will depend on the total number of different amplicons represented. As a starting point, we have found that a theoretical amount of 100,000 reads per amplicon provides good quantification ratios. A typical targeted-DAP assay may consist of the binding assay of one protein of interest to a DNA pool comprising approximately 20 different amplicons, covering both wild-type and mutant versions. In addition, it is recommended to test several amplification conditions (e.g., 10, 12, 14, 16, 18, and 20 cycles), as well as one input sample to quantify the representation of each amplicon within the sample, making 7 libraries in total. Therefore, a read number of 14 M would generate good quality data for relative quantification of the binding to DNA of a protein of interest, what can be easily achieved in a MiSeq Sequencing System using a MiSeq Reagent Kit v2.

Acknowledgments

This work has been funded by grants from the Spanish Ministry of Economy (MINECO) and fondos FEDER (PID2020-119451GB-I00 and PID2023-149400NB-I00). J.M.F.Z. acknowledges financial support from the Spanish State Research Agency, AEI/10.13039/501100011033, through the "Severo Ochoa" Programme for Centres of Excellence in R&D (SEV-2017-0712 and CEX2023-001386-S). E.A. is a FPI-SO fellow funded by MCIU/AEI (PRE2020-093525).

Conflict of Interest The authors have no conflicts of interest to declare that are relevant to the content of this chapter.

References

1. Meyer RS, Purugganan MD (2013) Evolution of crop species: genetics of domestication and diversification. Nat Rev Genet 1412(14): 840–852. https://doi.org/10.1038/nrg3605
2. Swinnen G, Goossens A, Pauwels L (2016) Lessons from domestication: targeting cis-regulatory elements for crop improvement. Trends Plant Sci 21:506–515. https://doi.org/10.1016/j.tplants.2016.01.014
3. Bartlett A, O'Malley RC, Huang SC et al (2017) Mapping genome-wide transcription-factor binding sites using DAP-seq. Nat Protoc 12:1659–1672. https://doi.org/10.1038/nprot.2017.055
4. O'Malley RC, Huang SC, Song L et al (2016) Cistrome and epicistrome features shape the regulatory DNA landscape. Cell 165:1280–1292. https://doi.org/10.1016/j.cell.2016.08.063
5. Franco-Zorrilla JM, López-Vidriero I, Carrasco JL et al (2014) DNA-binding specificities of plant transcription factors and their potential to define target genes. Proc Natl Acad Sci 111: 2367–2372. https://doi.org/10.1073/pnas.1316278111
6. Godoy M, Franco-Zorrilla JMJM, Pérez-Pérez J et al (2011) Improved protein-binding microarrays for the identification of DNA-binding specificities of transcription factors. Plant J 66:700–711. https://doi.org/10.1111/j.1365-313X.2011.04519.x
7. Franco-Zorrilla JM, Solano R (2014) High-throughput analysis of protein-DNA binding affinity. In: Methods in molecular biology. Humana Press, pp 697–709. https://doi.org/10.1007/978-1-62703-580-4_36
8. Franco-Zorrilla JM, Prat S (2021) DAP-Seq identification of transcription factor-binding sites in potato. Methods Mol Biol 2354:123–142. https://doi.org/10.1007/978-1-0716-1609-3_6
9. Kapust RB, Waugh DS (1999) Escherichia coli maltose-binding protein is uncommonly effective at promoting the solubility of polypeptides to which it is fused. Protein Sci 8:1668–1674. https://doi.org/10.1110/ps.8.8.1668
10. López-Vidriero I, Godoy M, Grau J et al (2021) DNA features beyond the transcription factor binding site specify target recognition by plant MYC2-related bHLH proteins. Plant Commun 2:100232. https://doi.org/10.1016/j.xplc.2021.100232
11. Fernández-Calvo P, Chini A, Fernández-Barbero G et al (2011) The Arabidopsis bHLH transcription factors MYC3 and MYC4 are targets of JAZ repressors and act additively with MYC2 in the activation of jasmonate responses. Plant Cell 23:701–715. https://doi.org/10.1105/tpc.110.080788
12. Hufford MB, Xu X, Van Heerwaarden J et al (2012) Comparative population genomics of maize domestication and improvement. Nat Genet 44:808–811. https://doi.org/10.1038/ng.2309
13. Li X, Schmitz RJ (2025) Cis-regulatory dynamics in plant domestication. Trends Genet. https://doi.org/10.1016/J.TIG.2025.02.005
14. Schmitz RJ, Grotewold E, Stam M (2022) Cis-regulatory sequences in plants: their importance, discovery, and future challenges. Plant Cell 34:718–741. https://doi.org/10.1093/plcell/koab281
15. Rodríguez-Leal D, Lemmon ZH, Man J et al (2017) Engineering quantitative trait variation for crop improvement by genome editing. Cell 171:470–480.e8. https://doi.org/10.1016/j.cell.2017.08.030
16. Wang X, Aguirre L, Rodríguez-Leal D et al (2021) Dissecting cis-regulatory control of quantitative trait variation in a plant stem cell circuit. Nat Plants 7:419–427. https://doi.org/10.1038/s41477-021-00898-x
17. Liu L, Gallagher J, Arevalo ED et al (2021) Enhancing grain-yield-related traits by CRISPR–Cas9 promoter editing of maize CLE genes. Nat Plants 7:287–294. https://doi.org/10.1038/S41477-021-00858-5
18. Chen K, Wang Y, Zhang R et al (2019) CRISPR/Cas genome editing and precision plant breeding in agriculture. Annu Rev Plant Biol 70:667–697. https://doi.org/10.1146/annurev-arplant-050718

Chapter 8

The Yeast Two-Hybrid System: In Vivo Detection of Direct Protein-Protein Interactions

Pablo Morales-Martínez and Araceli G. Castillo

Abstract

Yeast two-hybrid systems are one of the most popular and cost-effective in vivo genetic approaches for screening protein-protein interactions. The rationale behind the yeast two-hybrid (Y2H) system is the modular nature of transcription factors, which typically consist of a DNA-binding domain (BD) and an activation domain (AD), and the finding that both domains are not required to be covalently linked to activate transcription. By splitting a transcription factor into two separate domains, one fused to a "bait" protein (BD fusion) and the other to a "prey" protein (AD fusion), interaction between the bait and prey proteins brings the two domains together. This reconstitution of the transcription factor results in the activation of a reporter gene, signaling the protein-protein (bait-prey) interaction. Several variants of the Y2H system exist, but herein we will focus on the one based on the GAL4 transcription factor. In this chapter, we provide detailed protocols to efficiently co-transform and store competent yeast cells, test the interaction between two individual proteins, measure the relative strength of that interaction, and check the expression of the fusion proteins.

Key words Yeast two-hybrid system, Y2H, Protein-protein interaction, GAL4, Bait protein, Prey protein, Interactors

1 Introduction

The yeast two-hybrid (Y2H) system is a molecular biology technique used to study in vivo direct protein-protein interactions in yeast cells, specifically *Saccharomyces cerevisiae* [1]. The Y2H exploits the modular nature of many site-specific transcriptional activators, which consist of a DNA-binding domain (BD) and a transcriptional activation domain (AD), as well as the fact that the two domains can be brought together by the interaction of any two proteins, therefore activating gene transcription even if these domains are not covalently linked (Fig. 1). To determine whether two proteins interact, both must be expressed fused to either a BD or an AD in a yeast strain that contains one or more reporter genes with upstream binding sites for the DNA-binding domain. If both

José M. Franco-Zorrilla and Carmen Martín-Pizarro (eds.), *Plant Transcription Factors: Methods and Protocols*, Methods in Molecular Biology, vol. 2985, https://doi.org/10.1007/978-1-0716-4972-5_8,

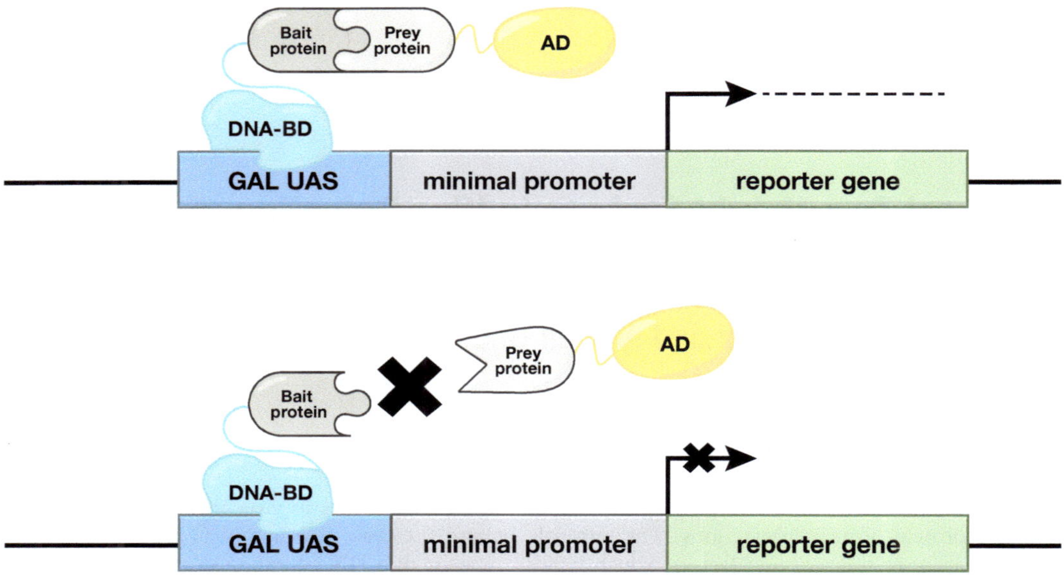

Fig. 1 GAL4 yeast two-hybrid system. The DNA-BD of the yeast GAL4 protein fused to the bait protein binds the GAL UAS upstream of the reported gene. The AD of the GAL4 protein fused to the prey protein functions as a transcriptional activator if the interaction between the bait and the prey proteins occurs

hybrid proteins interact, they create a functional activator by bringing the activation domain in proximity with the DNA-binding domain, which will finally lead to the transcription of the reporter gene (Fig. 1). This technology can also be used to identify novel protein-protein interactions using a bait protein fused to BD (bait protein) and a plasmid library that expresses cDNA-encoded proteins fused to the AD transcription activation domain (prey proteins). This latter application will not be presented in this chapter, but excellent reviews can be found elsewhere [2, 3].

Yeast two-hybrid system has constituted a powerful tool in molecular biology since its development in the 1980s and has greatly benefited from the development and application of the Gateway cloning system [4]. One of its main advantages is that Y2H technology provides a sensitive method for detecting relatively weak and transient protein interactions. Moreover, since the interactions occur inside living yeast cells, Y2H can provide information about the interactions in a more physiological context than in vitro assays. Finally, Y2H is relatively straightforward to set up and implement compared to other methods like co-immunoprecipitation or mass spectrometry-based approaches, and it does not require expensive reagents or specialized equipment. Despite its advantages, the Y2H system has also some limitations that must be considered prior use.

- *Dependence on the fusion protein functionality:* The proteins of interest are fused to the domains (BD and AD) of a transcription factor. If these fusions interfere with the protein's native function or localization, they can affect the interaction results.
- *Yeast-specific limitations*: Not all proteins are expressed, folded, or posttranslationally modified in yeast as they are in their native organism, which can affect the detection of interactions.
- *Interactions that require a specific cellular context*: Some interactions may depend on specific cellular components or conditions not present in yeast, limiting the applicability of Y2H for certain protein interactions.
- *Limited to binary interactions*: Y2H is primarily designed to detect interactions between two proteins at a time, making it less effective for studying complex protein networks involving multiple proteins simultaneously.

Overall, while the Y2H system is a valuable tool for exploring protein interactions, researchers must carefully weigh up its limitations and complement it with other methods to confirm and validate the newly identified protein-protein interactions.

Several varieties of the Y2H systems exist, commonly involving DNA-binding domains that derive from the yeast Gal4 protein or the *Escherichia coli* LexA protein [5, 6]. Transcriptional activation domains are commonly derived from the Gal4 protein, the herpes simplex virus VP16 protein, or a protein encoded by random *E. coli* sequences, the B42 "acid blob." Reporter genes include the *E. coli lacZ* gene and selectable yeast genes such as *HIS3* and *ADE2* (GAL4 system) or *LEU2* (LexA system). In this chapter, we focus on the Y2H system based on the yeast Gal4 protein where a bait gene is expressed as a fusion to the GAL4 DNA-binding domain (DNA-BD, bait protein), while another gene or cDNA is expressed as a fusion to the GAL4 activation domain (AD, prey protein) [5, 7]. We provide detailed protocols for the yeast transformation, storage of yeast competent cells, detection of the fusion protein expression, selection of the positive interactors discarding the autoactivating proteins, and the measurement of the interactions to assay their relative strength.

2 Materials

2.1 Plasmids

Many different DNA-binding domain (BD) and transcription activation domain (AD) containing vectors have been successfully used in yeast two-hybrid experiments. In many of these vectors, bait and prey inserts are expressed as GAL4 fusions with c-Myc and hemagglutinin (HA) epitope tags, respectively, which allows the detection of expression of both fusion proteins (Table 1). Otherwise, anti-

Table 1
Common bait and prey vectors used for the GAL4-based Y2H system

Plasmid	Fusion	Epitope	Yeast selection	Bacterial selection	Map/description
pGBKT7	BD/bait	c-Myc	*TRP1*	Kanamycin	Clontech protocol no. PT3024-1
pGBT9	BD/bait	None	*TRP1*	Ampicillin	Clontech protocol no. PT3062-1
pAS2–1	BD/bait	None	*TRP1*	Ampicillin	Clontech protocol no. PT3062-1
pGADT7	AD/prey	HA	*LEU2*	Ampicillin	Clontech protocol no. PT3024-1
pACT2	AD/prey	HA	*LEU2*	Ampicillin	Clontech protocol no. PT3062–1

GAL4 BD and AD monoclonal antibodies can be used (Subheading 3.9). In our experience, pGBKT7 and pGADT7 give the best results, and there are versions available that use the Gateway Recombinational Cloning [4]. Plasmid information of the most used GAL4-based Y2H is provided in Table 1.

2.2 Yeast Strains

Several yeast strains containing different reporter genes are available. Among them, PJ69-4A and AH109 strains grow consistently, especially when provided with supplemental adenine in complete media (up to 100 mg/L), and are frequently used. They both include three reporter genes—*ADE2*, *HIS3*, and *lacZ*—under the control of distinct GAL4 upstream activating sequences (UASs) and a minimal promoter (TATA box). These promoters yield strong and specific responses to Gal4 transcription factor. Moreover, the host strain AH109 contains a fourth reporter gene, *MEL1*, which is an endogenous gene found in several yeast strains that encodes a secreted α-galactosidase [8].

The *ADE2* reporter alone provides strong nutritional selection; however, the addition of *HIS3* selection reduces the incidence of false positives and allows to control the stringency of the selection (detailed in Subheading 2.3) [9]. Moreover, the use of the *lacZ* reporter gene, which encodes a β-galactosidase, allows to perform traditional β-galactosidase studies from yeast growing on plates (β-galactosidase assay on filter; Subheading 3.6) but also address the relative strength of the interactions using a quantitative β-galactosidase liquid assay (Subheading 3.7). Moreover, in response to GAL4 activation, the host strain AH109 will secrete a α-galactosidase from the *MEL1* gene, which can be detected on medium containing X-α-Gal [8].

The phenotype and genotype of the PJ69-4A and AH109 strains are shown in Table 2, and the protocols presented in this chapter can be used with both strains.

Table 2
Genotype and phenotypic characteristics of PJ69-4A and AH109 yeast strains

Strain	Genotype	Transformation markers	Interaction reporters	References
PJ69-4A	*MATa, trp1-901, leu2-3, 112, ura3-52, his3-200, gal4Δ, gal80Δ, LYS2::GAL1*$_{UAS}$-*GAL1*$_{TATA}$-*HIS3, GAL2*$_{UAS}$-*GAL2*$_{TATA}$-*ADE2*[a] *MET2:: GAL7*$_{UAS}$-*GAL7*$_{TATA}$-*LacZ*	*trp1, leu2, ura3*	*HIS3, ADE2, lacZ*[b]	
AH109	*MATa, trp1-901, leu2-3, 112, ura3-52, his3-200, gal4Δ, gal80Δ, LYS2::GAL1*$_{UAS}$-*GAL1*$_{TATA}$-*HIS3, GAL2*$_{UAS}$-*GAL2*$_{TATA}$-*ADE2*[a], *URA3::MEL1*$_{UAS}$-*MEL1*$_{TATA}$-*lacZ, MEL1*$_{UAS}$-*MEL1*$_{TATA}$-*MEL1*	*trp1, leu2, ura3*	*HIS3, ADE2, lacZ*[b], *MEL1*[c]	[9] A. Holtz, unpublished

The *trp1, his3, gal4*, and *gal80* mutations are deletions; *leu2-3, 112* allele is a double mutant. The *HIS3, ADE2*, and *lacZ* reporter genes are under the control of three heterologous GAL4-responsive UAS and promoter elements, GAL1, GAL2, and GAL7/MEL1, respectively

[a]The *ade2-101* gene of the precursor strain was replaced (by recombination) with the GAL2-ADE2 reporter construct

[b]*lacZ* reporter gene, i.e., is positive for β-galactosidase activity

[c]*MEL1* reporter gene, i.e., is positive for α-galactosidase activity

2.3 Media and Supplements

Prepare all media using deionized water. For solutions, use ultrapure water and analytical grade reagents. All reagents must be prepared and stored at room temperature (unless otherwise indicated).

1. *Yeast extract peptone dextrose adenine (YPDA):* 10 g/L yeast extract, 20 g/L peptone, 20 g/L glucose, and 20 mg/L adenine hemisulfate salt. Autoclave. If preparing solid YPDA, add agar (20 g/L) and then autoclave it (121 °C, 15 min).
2. *Minimal media (MM):* 1.7 g/L Yeast Nitrogen Base (YNB), 5 g/L ammonium sulfate, and 20 g/L glucose. Adjust pH to 6.5, and then autoclave (121 °C, 15 min). If preparing solid MM, add 20 g/L agar and then autoclave it. YNB is a base medium for preparation of minimal and synthetic defined yeast media that can be purchased from different companies, but it is essential that it does not contain amino acids. Amino acid solutions must be sterilized separately and then added to the sterilized media before pouring plates (Tables 3 and 4).
3. *Co-transformation media (MM4)*: MM with supplements but not leucine and tryptophan (Table 3).
4. *Interaction media (MM3, MM2, MM1, MMZ):* MM with different combination of supplements and amino acids (Table 3).
5. *3-Amino-1,2,4-triazole* (3-AT; Sigma-Aldrich): A 2 M stock solution should be freshly prepared by dissolving 3-AT in ultrapure H_2O and filter-sterilized. Add to the autoclaved

Table 3
Supplements added to MM to produce the co-transformation and the interaction media

		Interaction media			
Supplements	**Co-transf. Media MM4**	**Medium stringency MM3**	**Medium stringency MM2**	**High stringency MM1**	**High stringency MMZ[b]**
Leucine	–	–	–	–	–
Tryptophan	–	–	–	–	–
Lysine	+	+	+	+	+
Methionine	+	+	+	+	+
Uracil	+	+	+	+	+
Adenine	+	+	–	–	–
Histidine	+	–	+	–	–
3-AT[a]	–	+	–	+	+
X-α-Gal[b]	–	–	–	–	+

[a]3-Amino-1,2,4-triazole (3-AT; Sigma-Aldrich Cat. No. A-8056) is a competitive inhibitor of the *HIS3*-gene product and may be used to titrate the minimum level of *HIS3* expression required for growth on histidine-deficient media
[b]Just use with AH109 strain. To select in plates for the expression of the MEL1 reporter gene

Table 4
Recommended final concentration of supplements in MM

Supplements	Final concentration	Stock solution	Add to 1 L of media
L-Adenine hemisulfate salt	20 mg/L	2 g/L	10 mL
L-Histidine HCl monohydrate	20 mg/L	10 g/L	2 mL
L-Leucine	100 mg/L	10 g/L	10 mL
L-Lysine HCl	50 mg/L	10 g/L	5 mL
L-Methionine	20 mg/L	10 g/L	2 mL
L-Tryptophan[a]	20 mg/L	10 g/L	2 mL
L-Uracil	20 mg/L	2 g/L	10 mL
3-AT[b]	2 mM	2 M	1 mL
X-α-Gal[c]	20 mg/L	20 mg/mL	1 mL

Stock solutions may be autoclaved (121 °C, 15 min) or filter-sterilized using a 0.2 μm filter and stored at room temperature or at 4 °C for up to 1 year
[a]Stock must be protected from light, filtered, and kept at 4 °C
[b]Stock must be protected from light, filtered, and used fresh
[c]Store X-α-Gal solutions in glass or polypropylene bottles at –20 °C and must be protected from light

agar medium (cool to 55 °C before adding it). Please be aware that different yeast two-hybrid strains need different 3-AT concentrations in the media.

6. *X-α-Gal (5-Bromo-4-chloro-3-indolyl α-D-galactopyranoside)*: Dissolve X-α-Gal at 20 mg/mL in dimethylformamide (DMF). Store X-α-Gal solutions in glass or polypropylene bottles at −20 °C, protected from light. When adding to autoclaved agar medium, ensure the solution is cooled to 55 °C before mixing.

   The media requirements for PJ69-4A strain are summarized in Table 3 (the same supplements are needed for the AH109 strain, but the addition of methionine is not necessary). The recommended final concentration of the supplements is shown in Table 4.

2.4 Yeast Transformation

Solutions and reagents required for yeast transformation are the following ones (use ultrapure water for all solutions and analytical grade reagents):

1. 50% PEG4000 (Polyethylene glycol, avg. MW = 3350). Filter-sterilized or autoclaved.
2. 1 M lithium acetate; adjust pH to 7.5 using diluted acetic acid, and filter-sterilized.

2.5 β-Galactosidase/lacZ Assay on Filter/Colony-Lift Filter Assay

1. Sterile filter paper (Whatman #5 or VWR Grade 410 paper filters). Nitrocellulose filters also can be used, but they are prone to crack when frozen.
2. Liquid nitrogen (it should be handled with care; always wear thick gloves and goggles).
3. Z buffer (adjust to pH 7.0 and autoclave). Can be stored at room temperature.

   16.1 g/L $Na_2HPO_4 \cdot 7H_2O$

   5.5 g/L $NaH_2PO_4 \cdot H_2O$

   0.75 g/L KCl

   0.246 g/L $MgSO_4 \cdot 7H_2O$.
4. X-gal (5-bromo-4-chloro-3-indolyl-β-D-galactopyranoside) stock solution: 20 mg/mL X-Gal in N,N-dimethylformamide (DMF). Store in the dark at −20 °C.
5. Z buffer/X-gal solution.

   100 mL Z buffer

   0.27 mL β-mercaptoethanol

   1.67 mL X-gal stock solution.

2.6 Liquid β-Galactosidase/lacZ Assay

1. Z buffer (as in Subheading 2.5): 0.1 M sodium phosphate monobasic pH 7, 10 mM potassium chloride, 1 mM magnesium sulphate. Adjust pH to 7.0 and autoclave. Prepare fresh before each use.
2. o-Nitrophenyl β-D-galactopyranoside (ONPG): Prepare solution fresh before each use at 4 mg/mL in Z buffer. Adjust to pH 7.0 and mix well. ONPG requires 1–2 h to dissolve.
3. 1 M Na_2CO_3 in deionized H_2O (to stop the reaction).

2.7 Liquid α-Galactosidase Assays

1. PNP-α-Gal solution (prepare solution fresh before each use): 100 mM (p-nitrophenyl α-D-Galactopyranoside) in deionized H_2O. Filter- sterilized.
2. Assay buffer: Two volumes of 0.5 M sodium acetate (pH 4.5) with 1 volume PNP-α-Gal solution [2:1 (v/v) ratio]. Mix well. Prepare fresh before each use.
3. 0.1 M Na_2CO_3 in deionized H_2O (to stop the reaction).

2.8 Protein Extraction from Yeast

1. *Acid-washed glass beads* (425–600 μm diameter, Sigma-Aldrich).
2. FastPrepTM FP120 (BIO 101) to grind the cells.
3. *RIPA buffer*. 2 mM sodium phosphate buffer pH 7, 0.2% Triton X-100, 0.02% SDS, 0.2 mM EDTA pH 8, and 10 mM NaCl. Store at room temperature.
4. *Protease inhibitor cocktail* (Sigma-Aldrich).
5. *100 mM phenylmethylsulfonyl fluoride, PMSF*, prepared in methanol (Sigma-Aldrich).
6. *2× Laemmli sample buffer:* 10% SDS, 20% glycerol, 120 mM Tris-HCl pH 6.8, and 0.02% bromophenol blue. Store at room temperature.

2.9 Antibodies to Detect the Fusion Proteins

1. Bait proteins: Gal4 DNA-BD monoclonal antibody or anti-Myc monoclonal antibody (it will depend on the Y2H vector used; see Table 1).
2. Prey proteins: Gal4 DNA-AD monoclonal antibody or anti-HA monoclonal antibody (it will depend on the Y2H vector used; see Table 1).

3 Methods

3.1 Cloning the Full-Length or Partial cDNA from the Proteins of Interest

To determine if two proteins (A and B) interact, a complete or partial cDNA from each protein must be cloned into both bait and prey vectors to express them fused either to the BD or the AD of GAL4, respectively. The following proceedings must be taken into consideration:

(a) For the generation of a specific hybrid, the gene encoding the protein A (or B) is inserted into the multiple cloning site of the BD- and AD-vectors (pGBKT7 and pGADT7 for example) in the correct reading frame such that a hybrid protein is generated with either the BD-GAL4 or the AD-GAL4.

(b) It is recommended to generate both hybrids for each protein, cloning the gene for protein A (or B) in both the bait (BD) and the prey (AD) vectors (Table 1). This will allow to test the interaction between two proteins in both directions (BD-A and AD-B but also BD-B and AD-A), and also test if each protein dimerizes or self-interact (Tables 5 and 6).

(c) It is essential to generate a bait and a prey vector whose fusion proteins interact (positive control for the Y2H) and a bait and a prey vector whose fusion proteins do not interact (negative control for the Y2H). BD-p53 (murine p53 fragment from aa 66 to 230, GenBank Accession #K01700), BD-lamC (human laminin C fragment from aa 72 to 390, GenBank Accession #M13451), and AD-AgT (SV40 large T-antigen insert from aa 87 to 708, GenBank Locus SV4CG) are the most widely used positive and negative controls in the GAL4 Y2H assays (Table 5 and Table 6). For sequences and maps from these control fragments, see Clontech Protocol No. PT3062-1.

Table 5
Number of co-transformants (Co-T, numbered 1-10) needed to determine if two proteins, A and B, interact by Y2H

Bait/Prey	AD-A	AD-B	AD-AgT
BD-A	Co-T3	Co-T1	Co-T5
BD-B	Co-T2	Co-T4	Co-T6
BD-lamC	Co-T7	Co-T8	Co-T9
BD-p53	–	–	Co-T10

A and B: two proteins of interest
LamC laminine C, *AgT* SV40 large T-antigen, *p53* murine p53

Table 6
Analysis of the results if the Y2H co-transformants (Co-T) grow one or more of the different interaction media (MM3, MM2, and MM1)

Co-T	Conclusion	Comment
Co-T1	BD-A and AD-B interact	There should be no growth of Co-T5, Co-T6, Co-T7, Co-T8
Co-T2	BD-A and AD-B interact	There should be no growth of Co-T5, Co-T6, Co-T7, Co-T8
Co-T3	Protein A self-interact	May function as a dimer
Co-T4	Protein B self-interact	May function as a dimer
Co-T5	BD-A autoactivates	Other GAL4-AD fusion proteins should be tested
Co-T6	BD-B autoactivates	Other GAL4-AD fusion proteins should be tested
Co-T7	AD-A autoactivates	Other GAL4-BD fusion proteins should be tested
Co-T8	AD-B autoactivates	Other GAL4-BD fusion proteins should be tested
Co-T9	No growth on these media[a]	Negative control for the GAL4 Y2H system
Co-T10	Growth on every media[b]	Positive control for the GAL4 Y2H system

All the Co-T should grow on the co-transformation media, MM4
[a]Yeast will be colorless on β-galactosidase (Colony-lift Filter Assay) or α-galactosidase assays (MMZ)
[b]Yeast will be blue on β-galactosidase Colony-lift Filter Assay or α-galactosidase assays (MMZ)

3.2 Designing a Y2H Assay

Protein-protein interactions are tested by co-transforming the yeast cells with the bait and prey vectors according to the protocol detailed in Subheadings 3.3 and 3.4. However, protein expression and putative autoactivation of bait and prey proteins must also be tested in every Y2H experiment. The specificity of the interaction must be also confirmed using bait or prey vectors expressing non-related proteins. Therefore, it is advisable to co-transform not just the hybrid proteins whose interaction is going to be tested but also perform other co-transformations that will determine the robustness and the specificity of the interaction. Table 5 summarizes the co-transformations required in a Y2H experimental design. The information given by the co-transformants once they are placed on interaction media is indicated in Table 6.

3.3 Preparation of Yeast Competent Cells

All the steps described below must be performed using a Bunsen burner or in a laminar flow cabinet.

1. Streak an YPDA plate with cells from the −80 °C yeast stock. Incubate the plate upside down at 30 °C until colonies appear (~3 days) (*see* **Note 1**). Yeast strains can be stored for up to 1 month at 4 °C on YPDA medium in petri dishes sealed with parafilm. However, fresh colonies will give better results when inoculating a liquid culture.

2. Inoculate one colony into 20 mL of YPDA medium in a sterile 50 mL flask. Incubate overnight at 30 °C with shaking at 250 rpm.
3. Transfer an appropriate volume of the overnight culture to 50 mL of YPDA in a 250 mL flask to reach an $O.D_{600}$ of 0.2. Incubate shaking at 30 °C (250 rpm) until the $O.D_{600}$ reaches 0.8–1 (5–7 h). 50 mL of culture will allow you to perform 20 co-transformations. Scale up if necessary.
4. Centrifuge the cells at 3000 × g for 4 min at room temperature. Discard the supernatant and resuspend the pellet in 25 mL of sterile deionized water.
5. Centrifuge the cells at 3000 × g for 4 min at room temperature. Discard the supernatant and resuspend the pellet in 25 mL of 100 mM lithium acetate.
6. Centrifuge the cells at 3000 × g for 4 min at room temperature. Discard the supernatant and resuspend the pellet in 800–900 μL of 100 mM lithium acetate (final volume should be 1 mL). Transfer the cell suspensions to a 1.5 mL microcentrifuge tube. The cells are now competent cells and are ready to be transformed with plasmid DNA.
7. Optional: Competent yeast cells can be frozen and stored for up to 1 year at −80 °C with little loss of transformation efficiency. Centrifuge the cells at 3000 × g for 4 min at room temperature. Discard the supernatant and resuspend the pellet in 350–400 μL of filtered sterile frozen competent cell solution (10% v/v high quality DMSO, 5% v/v glycerol). The final volume should be 0.5 mL. Make 50 μL aliquots in sterile 1.5 mL microcentrifuge tubes, and place them into a tube Styrofoam rack. Place the rack in a box (Styrofoam or cardboard) with additional insulation such as tissue paper to reduce the air space around the sample box. This will result in the samples freezing slowly, which is essential for good survival rates. Put the box in a −80 °C freezer overnight. The frozen yeast cells can then be removed from the freezing container and stored at −80 °C.

3.4 Yeast Transformation

1. For each co-transformation, use 50 μL of competent cells and add in the indicated order: 240 μL of 50% PEG, 36 μL of 1 M lithium acetate, and the transforming DNA (co-transformation with the bait and prey vectors as indicated in Table 5) dissolved in deionized sterile water (200–500 ng of each plasmid in a final volume of 75 μL). Salmon sperm DNA is just added when high transformation efficiency is needed (i.e., library transformation) (*see* **Note 2**). To test the interaction between two proteins, A and B, a minimum of ten co-transformations should be performed as indicated in Table 5. In that case,

prepare ten 1.5 mL microcentrifuge tubes with the plasmid DNA (200–500 ng of each plasmid), one tube with no DNA (negative control for the transformation), and add 50 μL of competent cells to each tube and leave them at room temperature. Prepare in a sterile 50 mL tube the following master mix: 2880 μL of 50% PEG (240 μL × 12), 432 μL of 1 M lithium acetate (36 μL × 12), and deionized sterile water ([75 μL – volume from the plasmid DNA = 71 μL aprox.] 71 μL × 12). Vortex the master mix vigorously, and add 347 μL to each microcentrifuge tube that already contains the competent yeast cells and the DNA.

2. Vortex each 1.5 mL microcentrifuge tube and incubate at 30 °C for 30 min.
3. Mix by gently inverting the tubes and transfer to a preheated 42 °C circulating water bath or thermoblock. Incubate for 30 min.
4. Allow the cells to cool at room temperature for 1–2 min.
5. Centrifuge the cells at 3000 × g for 3 min at room temperature. Discard all the supernatant with a p.1000, and resuspend the pellet in 1000 μL of deionized sterile water using the p.1000.
6. Spread 100 μL and 900 μL (after centrifugation) of the resuspended cells onto MM4 co-transformation media plates (10 cm of diameter).
7. Invert the plates and incubate at 30 °C for 3–5 days.

3.5 Detect the Protein-Protein Interaction in Yeast Cells

1. Select at least two colonies for each co-transformation and, with a sterile toothpick, do several X-streaks for each colony on MM4 plates (Fig. 2). If the colonies are closely spaced, it will be necessary to streak purify to single colonies to separate the different co-transformants. Incubate the plates at 30 °C for 2 days.
2. Using a multichannel pipette, do 1/10 serial dilutions with deionized sterile water in a 96-well sterile plate. Add 200 μL of deionized sterile water to the first column and 180 μL to the following five columns.
3. Collect the yeast from one X-streak using a sterile toothpick and resuspend them in the 200 μL of deionized sterile water from the first column of the 96-well sterile plate. Once eight of the co-transformants have been resuspended on each row of the first column, do 1/10 serial dilutions (20 μL of cells on each well that contain 180 μL of water). On each 8 × 6 set, one colony from both Y2H controls (positive, BD-p53/AD-AgT, and negative, BD-lamC/AD-AgT) should be included (Fig. 2).

4. Use a stainless-steel replica plater (8×6) or a multichannel pipette (adding 5 μL of each cell suspension) (*see* **Note 3**) to do replicas on plates containing the following media (Fig. 2):
   (a) MM4 to check cell growth of the co-transformants. All cells should grow.
   (b) MM3 to select for the expression of HIS3. Growth indicates positive interaction (medium stringency).
   (c) MM2 to select for the expression of ADE2. Growth indicates positive interaction (medium stringency).
   (d) MM1 to select for the expression of HIS3 and ADE2. Growth indicates positive interaction (high stringency).
   (e) MMZ (just used with AH109 strain) to select for the expression of *HIS3*, *ADE2*, and *lacZ*. Growth and blue color indicate positive interaction (high stringency). If the interaction needs to be quantified, go to Subheading 3.7.
   (f) Colony-lift Filter Assay/β-galactosidase assay on filter (used for PJ69-4A strain; recommended when yeast strain contain one or only a few copies of the *lacZ* reporter gene). See Subheading 3.6.
5. Invert the plates and incubate them at 30 °C for 2–7 days. Check growth every day and scan/photo plates when necessary.
6. Positive interactions will be determined by comparing the growth of each co-transformant with the growth of the positive (BD-p53/AD-AgT) and the negative (BD-lamC/AD-AgT) controls on MM1, MM2, and MM3. Roughly, the same number of cells for each co-transformant should be noticeable on the MM4 plate. Taking into consideration the co-transformations (Co-T) performed on Table 5 and if all will grow similarly on MM4, we can consider that two proteins (A and B) interact if the Co-T1 and/or CoT2 grow on any on the interaction media assayed (MM3, MM2, and MM1), but Co-T5, Co-T6, Co-T7, and Co-T8 do not grow on any of them. The interpretation of the results from a Y2H assay is described on Table 6.

3.6 Colony-Lift Filter Assay/β-Galactosidase Assay on Filter

1. For best results, use fresh colonies. Therefore, select at least two colonies for each co-transformation, and with a sterile toothpick, do a X-streak/small patch for each colony on MM4 plates. Include as many colonies as you can fit in the MM4 plate, and always include the Y2H interaction controls, Co-T9 and Co-T10 (Table 5). Incubate the plates at 30 °C for 1–2 days.
2. Prepare Z buffer/X-gal solution (Subheading 2.5).

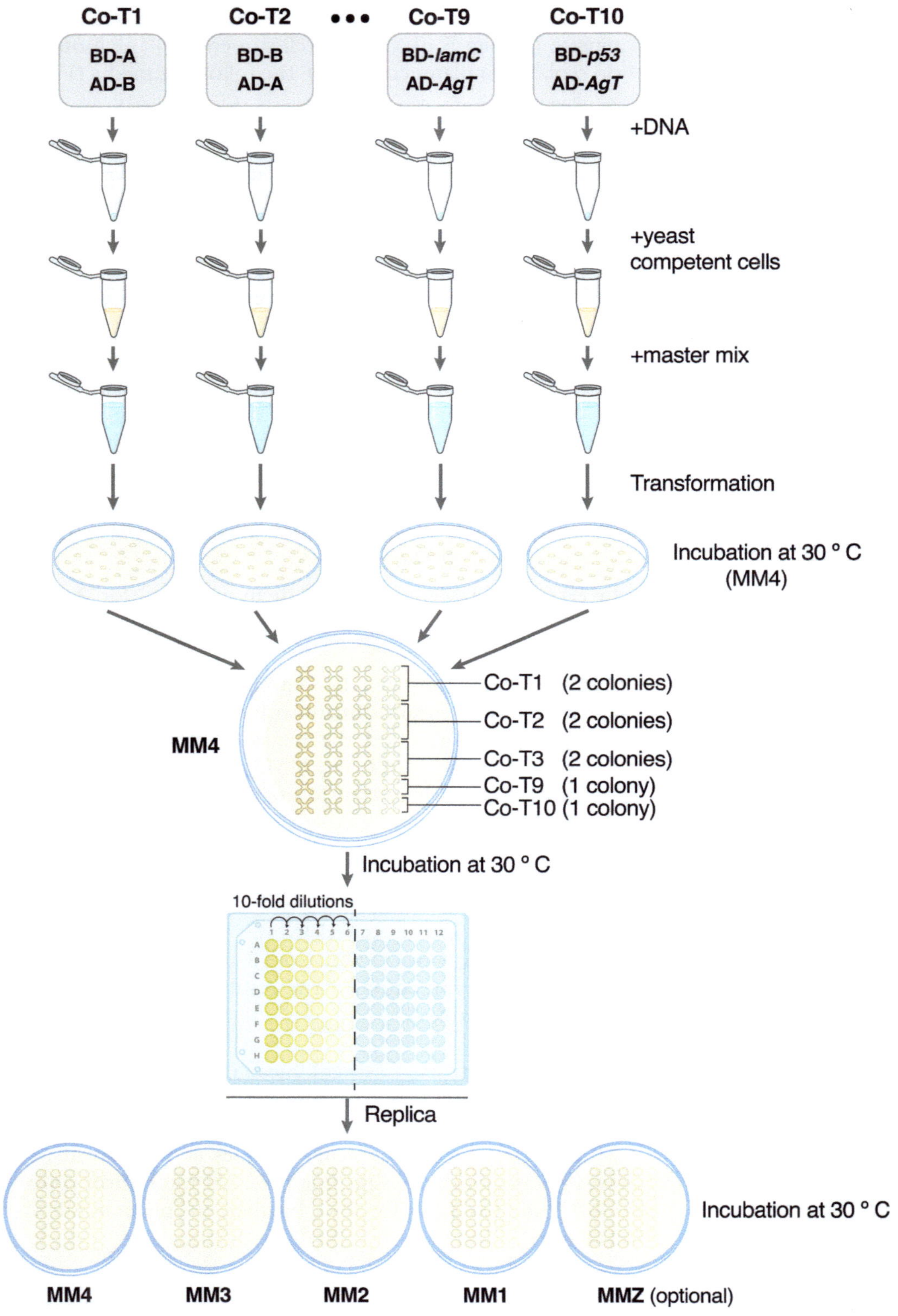

Fig. 2 Co-transformations and analysis of protein-protein interactions. 50 μL of competent yeast cells (on each 1.5 mL tube) are co-transformed with DB-bait and AD-prey constructs using the transformation master mix.

3. For each MM4 plate with the co-transformants to be assayed, presoak a sterile Whatman #5 or VWR grade 410 filter by placing it in 2.5–5 mL of Z buffer/X-gal solution in a clean 100- or 150-mm plate.
4. Using forceps, place a clean, dry filter over the surface of the MM4 plate with the co-transformants to be assayed. Gently rub the filter to help the biomass cling to the filter.
5. Carefully lift the filter with the co-transformants off the agar plate with forceps and transfer it (colonies facing up) to a pool of liquid nitrogen. Completely submerge the filter for 10 s.
6. After 10–15 s, remove the filter from the liquid nitrogen, leave it on a clean 100- or 150-mm plate, and allow it to thaw at room temperature. This freeze/thaw treatment permeabilizes the cells.
7. Carefully place the filter, colony side up, on the presoaked filter (from **step 3**). Avoid trapping air bubbles under or between the filters.
8. Incubate the filters at 30 °C (or room temperature) and check periodically for the appearance of blue colonies. The time it takes yeasts producing β-galactosidase to turn blue varies, typically from 30 min to 8 h (prolonged incubation may give false positives).

3.7 Measure the Protein-Protein Interaction: Liquid β-Galactosidase/lacZ Assay

To assay the relative strength of the interactions, a quantitative β-galactosidase assay could be used (*see* **Note 4**). Co-transformants must grow on liquid MM4 (no agar) which selects for the bait and the prey vectors. PJ69-4A exhibits a higher level of induced β-galactosidase activity than AH109 because of differences in the strengths of the lacZ promoter constructs (Table 2).

1. Pick three or four colonies for each co-transformant, and inoculate them in triplicate in 5–10 mL of liquid MM4 (all colonies should have similar growth). Incubate at 30 °C and 250 rpm until the $O.D_{600}$ reaches 0.5–0.8 (overnight, cells will grow slower than in YPDA). Annotate the $O.D_{600}$ from each culture. The positive and negative Y2H interaction controls should always be included (Tables 5 and 6).

Fig. 2 (continued) Co-transformants (Co-T1 … Co-T10) are plated in MM4 media plates and incubate at 30 °C for 3–5 days. Colonies grown in MM4 plates are selected (two per co-transformant) with a sterile toothpick to do several X-streaks for each colony on MM4 plates. Incubate 1–3 days at 30 °C. Do 1/10 serial dilutions with deionized sterile water in a 96-well sterile plate, and collect the yeast from one X-streak using a sterile toothpick, and resuspend them in 200 μL of deionized sterile water (first column). Do replicas on plates (MM4, MM3, MM2, MM1, and MMZ) and incubate at 30 °C for 2–7 days

2. Centrifuge at 3000 × g for 4 min at room temperature. Discard the supernatant and wash the pellet with Z buffer (Subheading 2.6). Centrifuge at 3000 × g for 4 min at room temperature, and resuspend the pellet in 300 μL of Z buffer.
3. Break the cells by subjecting the yeast suspension to three freeze/thaw cycles in liquid nitrogen.
4. Aliquots (100 μL) are then assayed for β-galactosidase activity by adding 700 μL of Z buffer containing 40 μL of β-mercaptoethanol and 160 μL of substrate o-Nitrophenyl-beta-D-Galactopyranoside (ONPG, 4 mg/mL). Add 400 μL of 1 M Na_2CO_3 to the mixture when the yellow color develops to stop the reaction.
5. Centrifuge at 12,000 × g for 10 min. Accumulation of the ONPG product is measured at 420 nm. Protein concentration is measured by the Bradford assay (Bio-Rad). The enzyme-specific activity is calculated as described [10]. The relative activities must be normalized against the maximum value obtained, which is set to 100%. The different interactions should be tested in a minimum of three experiments, each of which should assay three independent co-transformants for each interaction.

3.8 Measure the Protein-Protein Interaction: Liquid α-Galactosidase Assay

Co-transformants must grow on liquid MM4 (no agar) which selects for the bait and the prey vectors.

1. Pick three or four colonies for each co-transformant, and inoculate them in triplicate in 5–10 mL of liquid MM4 (all colonies should have similar growth). Incubate at 30 °C and 250 rpm until the $O.D_{600}$ reaches 0.5–0.8 (overnight, cells will grow slower than in YPDA). Annotate the $O.D_{600}$ from each culture. The positive and negative Y2H interaction controls should always be included (Tables 5 and 6).
2. Centrifuge at 10,000 × g for 2 min at room temperature until cells are completely pelleted. Transfer the supernatant to a clean tube.
3. Prepare enough "assay buffer" (Subheading 2.7). For each sample, 24 μL of "assay buffer" will be needed.
4. Transfer 8 μL of the supernatant from **step 2** to a 1.5 mL tube. Add 24 μL of "assay buffer," and incubate at 30 °C for 60 min (close properly to prevent evaporation).
5. Add 960 μL of 0.1 M Na_2CO_3 to the mixture to stop the reaction.
6. Record the optical density of each sample at 410 nm. The enzyme-specific activity is calculated as described [11]. The relative activities must be normalized against the maximum value obtained, which is set to 100%. The different interactions

should be tested in a minimum of three experiments, each of which should assay three independent co-transformants for each interaction.

3.9 Detecting the Expression of the Fusion Proteins

The expression of the fusion proteins (BD-A, AD-A, BD-B, and AD-B) should be checked specially if their interaction is not detected. For detection of the fusion proteins, a Western blot of protein extracts obtained from the co-transformed yeast could be performed using the following antibodies (Clontech Protocol No. PT1029-1):

1. Bait proteins: Gal4 DNA-BD monoclonal antibody or anti-Myc monoclonal antibody (it will depend on the Y2H vector used; see Table 1).
2. Prey proteins: Gal4 DNA-AD monoclonal antibody or anti-HA monoclonal antibody (it will depend on the Y2H vector used; see Table 1).

The protein extraction should be performed as follows:

1. Pick two independent colonies (from X-streaks on MM4 plates) for each co-transformant and inoculate them in 30–40 mL of liquid MM4. Incubate at 30 °C and 250 rpm until the OD_{600} reaches 0.8–1 (cells will grow slower than in YPDA). The positive and negative yeast two-hybrid interaction controls should always be included (Tables 5 and 6).
2. Centrifuge at 4000 × g for 3 min to pellet the cells. Wash the pellet once with cold water, transfer to an ice-cold screw-cap tube containing 0.5 mL of glass beads (425–600 μm), and resuspend in 200 μL of RIPA buffer containing protease inhibitor cocktail (50 μL to 5 mL of buffer) and 1 mM phenylmethylsulfonyl fluoride (PMSF). Grind in FastPrepTM FP120 (BIO 101) at power setting 5.5 for two-three 15-s intervals separated by 1-min intervals on ice.
3. Add 400 μL RIPA buffer with protease inhibitors and vigorously vortex. Keep on ice. Heat needle in a flame and quickly pierce bottom of the screw-cap tube containing lysate. Make two to three holes in each tube, and place this in a cold 1.5 mL tube. Centrifuge at 4000 × g for 1 min in cooled centrifuge (do six tubes at a time). Keep on ice.
4. Centrifuge at 4000 × g, 2 min in a cooled centrifuge. Carefully recover the filtered supernatant containing the protein extract. Protein concentration can be determined using a standard Bradford assay.
5. Set aside 50–100 μL of protein extract and add 2× Laemmli sample buffer. Load 100 μg of the protein extract on a polyacrylamide/SDS gel. The proteins can then be transferred to a

filter and blotted with standard immunoblotting (Western) methods.

For a faster but less efficient protocol to extract protein from yeast, *see* **Note 5**.

4 Notes

1. Incubation of the yeast cultures/plates can be done at 28 °C.
2. Minipreparation of plasmid DNA can be performed using the traditional lysis alkaline method as the RNA left in the extraction can work as a carrier increasing the transformation efficiency.
3. If using a stainless-steel replica plater to make the replica on each interaction plate, 8×6 dilutions can be assayed. If a multichannel pipette (adding 5 μL of each cell suspension) is used to do replicas on plates with the different interaction media, it is better to assay just 8×5 dilutions.
4. Quantitative liquid assays can be used to compare the relative strength of the protein-protein interactions observed in selected transformants. However, there is no direct correlation between β-galactosidase activity and the K_d of an interaction [12]. Furthermore, quantitative data cannot be compared between different host strains having different lacZ reporter constructs.
5. A faster but less efficient protocol to extract protein from yeast could be used:
   - Grow and recover the cells as described in Subheading 3.2.
   - Resuspend the pellet in 50 mL of 2× Laemmli sample buffer. The cells can then be broken by freezing on dry ice followed by boiling for 5 min prior to loading on an SDS polyacrylamide gel (about 15 μL/lane).

Acknowledgments

The authors would like to thank the GeminiTeamIHSM members for useful discussions.
This work was supported by the Spanish Ministerio de Economía y Competitividad (cofinanced by the European Regional Development Fund, ERDF) (grant PID2022-139376OB-C31 to A.G.C.) and by the Spanish Ministerio de Universidades (fellowship number FPU16/06137 to P.M.M.).

Conflict of Interest The authors have no conflicts of interest to declare that are relevant to the content of this chapter.

References

1. Fields S, Song O (1989) A novel genetic system to detect protein-protein interactions. Nature 340:245–246. https://doi.org/10.1038/340245a0
2. Reece-Hoyes JS, Walhout AJM (2018) High-efficiency yeast transformation. Cold Spring Harb Protoc 2018:563–568. https://doi.org/10.1101/pdb.prot094995
3. Reece-Hoyes JS, Walhout AJM (2018) Identifying interactors from an activation domain prey library. Cold Spring Harb Protoc 2018:553–562. https://doi.org/10.1101/pdb.prot094987
4. Reece-Hoyes JS, Walhout AJM (2018) Gateway recombinational cloning. Cold Spring Harb Protoc 2018:1–6. https://doi.org/10.1101/pdb.top094912
5. Fashena SJ, Serebriiskii IG, Golemis EA (2000) LexA-based two-hybrid systems. Methods Enzymol 328:14–26. https://doi.org/10.1016/s0076-6879(00)28387-0
6. Bartel P, Chien C-T, Sternglanz R, Fields S (1993) Using the two-hybrid system to detect protein-protein interactions. Cell Interact Dev A Pract Approach:153–180
7. Koegl M, Uetz P (2007) Improving yeast two-hybrid screening systems. Brief Funct Genomic Proteomic 6:302–312. https://doi.org/10.1093/bfgp/elm035
8. Aho S, Arffman A, Pummi T, Uitto J (1997) A novel reporter gene MEL1 for the yeast two-hybrid system. Anal Biochem 253:270–272. https://doi.org/10.1006/abio.1997.2394
9. James P, Halladay J, Craig EA (1996) Genomic libraries and a host strain designed for highly efficient two-hybrid selection in yeast. Genetics 144:1425–1436. https://doi.org/10.1093/genetics/144.4.1425
10. Chien CT, Bartel PL, Sternglanz R, Fields S (1991) The two-hybrid system: a method to identify and clone genes for proteins that interact with a protein of interest. Proc Natl Acad Sci USA 88:9578–9582. https://doi.org/10.1073/pnas.88.21.9578
11. Lazo PS, Ochoa AG, Gascón S (1978) Alpha-galactosidase (melibiase) from Saccharomyces carlsbergensis: structrual and kinetic properties. Arch Biochem Biophys 191:316–324. https://doi.org/10.1016/0003-9861(78)90094-2
12. Estojak J, Brent R, Golemis EA (1995) Correlation of two-hybrid affinity data with in vitro measurements. Mol Cell Biol 15:5820–5829. https://doi.org/10.1128/mcb.15.10.5820

Chapter 9

Identification of Protein-Protein Interactions in Strawberry Fruit Using Immunoprecipitation-Mass Spectrometry (IP-MS) Assay

Xiaojing Li and Bingbing Li

Abstract

Protein-protein interaction (PPI) serves as a fundamental regulatory mechanism that mediates diverse biological processes to govern cellular activities. The identification of protein-protein interactions is a pivotal technique for accurately analyzing protein function and systematically elucidating its regulatory mechanisms. Immunoprecipitation-mass spectrometry (IP-MS) represents a commonly employed approach for identifying protein-protein interactions in cells. Mass spectrometry enables the verification of both known protein interactions and the discovery of unknown proteins interacting with target proteins, providing not only information on the presence of proteins in samples but also facilitating comparative analysis of their relative abundance across different samples. In this chapter, we present a method for immunoprecipitating protein complexes from total plant protein extracts, followed by mass spectrometry (IP-MS), to identify the interaction between proteins in strawberry fruit.

Key words Protein-protein interaction, Immunoprecipitation, Mass spectrometry, Strawberry fruit

1 Introduction

Protein-protein interactions are fundamental to virtually all cellular processes, occurring in various cellular compartments and contributing to the regulation of biological functions. Understanding these interactions is crucial for unraveling the molecular mechanisms underlying cellular activities and signaling pathways [1–3]. Currently, protein-protein interactions can be detected using various methods such as yeast two-hybrid, bimolecular fluorescence complementation (BiFC), luciferase complementation imaging (LCI), Turbo ID, and protein immunoprecipitation followed by mass spectrometry (IP-MS). In this chapter, we present a comprehensive guide to the operational procedures and technical considerations involved in utilizing IP-MS for the identification of protein-protein interactions in strawberry fruits.

José M. Franco-Zorrilla and Carmen Martín-Pizarro (eds.), *Plant Transcription Factors: Methods and Protocols*, Methods in Molecular Biology, vol. 2985, https://doi.org/10.1007/978-1-0716-4972-5_9,

Immunoprecipitation (IP) using specific antibodies enables the isolation of target proteins and their associated interacting partners, providing a robust approach for investigating protein complexes and their interactions in vivo. IP combined with liquid chromatography-tandem mass spectrometry (IP-LC-MS) has emerged as a pivotal technique for exploring protein-protein interaction networks, particularly in plants [4, 5]. This strategy facilitates the comprehensive identification of protein complexes and posttranslationally modified proteins on a large scale, offering an extensive and intricate view of the cellular interactome [6–9]. IP-LC-MS is highly effective in detecting both transient and stable interactions, making it widely applicable in plant research. Its capacity to analyze complex protein mixtures and simultaneously identify numerous candidate interactors confers significant advantages over traditional methods. Given that plant systems often involve multifaceted signaling pathways, IP-LC-MS stands out due to its precision and scalability, thereby holding great potential for advancing our understanding of plant biology and posttranslational regulation.

2 Materials

Immunoprecipitation of protein complexes from strawberry fruit extracts can be achieved by utilizing an antibody that specifically targets either the protein of interest or an epitope tag, such as HA or FLAG, which is introduced into the protein sequence. The utilization of a tagged protein enables the use of commercially available antibodies with high specificity; however, this necessitates the establishment of transgenic lines expressing the tagged target protein (*see* **Note 1**). In this protocol, we provide a detailed description of immunoprecipitation using magnetic beads coated with anti-HA or anti-FLAG antibodies (*see* **Note 2**). If further optimization is conducted, alternative antibodies or bead types may be employed. All buffers and solutions are prepared using autoclaved Milli-Q water (prior to immunoprecipitation) or HPLC-grade water (post-immunoprecipitation). Appropriate waste disposal procedures should be followed for each reagent.

2.1 Tissue Collection and Preparation

1. Cryogenic tissue grinder.
2. Liquid nitrogen.
3. Low-temperature resistant gloves.
4. Nitrile gloves.
5. 2 mL centrifuge tubes.
6. Refrigerated centrifuge.

2.2 Protein Complex Isolation

1. Liquid nitrogen.
2. Nitrile gloves.
3. 2 mL centrifuge tubes.
4. 100 mM phosphate buffer, pH 7.8: For preparing 50 mL of phosphate buffer, pH 7.8, dissolve 1.432 g of disodium phosphate (Na_2HPO_4) and 0.156 g of monosodium phosphate (NaH_2PO_4) in 50 mL of Milli-Q water.
5. 20% Triton X-100: Prepare fresh in advance, allowing the Triton to mix with the Milli-Q on a roller for several hours.
6. 0.5 M EDTA, pH 8.0: Weigh 14.6 g of EDTA (disodium salt) and add it to a 1 L beaker; add approximately 900 mL of Milli-Q water; adjust the pH to 8.0 using a sodium hydroxide (NaOH) solution, as EDTA will only fully dissolve at this pH; once fully dissolved, bring the final volume to 1 L with Milli-Q water.
7. 100 mM phenylmethylsulfonyl fluoride (PMSF): Weigh 1.74 g of PMSF and dissolve it in 100 mL of isopropanol (or ethanol) in a fume hood, as PMSF is toxic and volatile. Mix thoroughly until fully dissolved.
8. 1 M Dithiothreitol (DTT): Weigh 15.43 g of DTT and dissolve it in 100 mL of Milli-Q water. Mix thoroughly until fully dissolved.
9. 10× protease inhibitor cocktail: Add one tablet per 10 mL of protein extract buffer (*see* **Note 3**).
10. 10× phosphatase inhibitor cocktail: Add one tablet per 10 mL of protein extract buffer.
11. Strawberry fruit protein extraction buffer: 100 mM phosphate buffer, pH 7.8, 1 mM EDTA, pH 8.0, 10% glycerol, 1% polyvinylpolypyrrolidone (PVPP), 0.5% Triton X-100, 1 mM DTT, 1 mM PMSF, 1× protease inhibitor cocktail, 1× phosphatase inhibitor cocktail.
12. Sonicator.

2.3 Protein Immunoprecipitation, Elution, and SDS-PAGE

1. Nitrile gloves.
2. Anti-HA/Flag magnetic microbeads.
3. Rotary mixer.
4. Refrigerated centrifuge.
5. 5× SDS loading buffer. Combine the following reagents in a 50 mL tube: 1.5 g SDS (sodium dodecyl sulfate), 2.5 mL 1 M Tris-HCl, pH 6.8, 2.5 mL glycerol, 500 μL β-mercaptoethanol (or 0.1 g DTT as an alternative), 2 mg bromophenol blue (for tracking dye). Add Milli-Q water to bring the final volume to

10 mL. Mix thoroughly until all components are dissolved. Aliquot and store at −20 °C.

6. 10% SDS-PAGE gel: 3.3 mL Milli-Q water, 2.5 mL 1.5 M Tris-HCl, pH 8.8, 4.0 mL 30% acrylamide/bis-acrylamide solution, 100 μL 10% SDS, 100 μL 10% ammonium persulfate (APS) and 4 μL tetramethylethylenediamine (TEMED) for 10 mL10% resolving gel; 3.4 mL Milli-Q water, 630 μL 1.0 M Tris-HCl, pH 6.8, 830 μL 30% acrylamide/bis-acrylamide solution, 50 μL 10% SDS, 50 μL 10% APS, and 5 μL TEMED for 5 mL 4% stacking gel.
7. 10× SDS-PAGE running buffer: 30.3 g Tris base, 144 g glycine, and 10 g SDS; add Milli-Q water to bring the final volume to 1 L.
8. Coomassie staining and destaining (10% acetic acid, 50% methanol in water) solutions.

2.4 LC-MS Equipment

NanoLC-LTQ-Orbitrap XL (Thermo Fisher Scientific).

3 Methods

Considering the involvement of protein-protein interactions in the IP-MS experiment, it is crucial to understand that the activity of extracted plant proteins can impact the final experimental outcomes. To ensure the reliability of results, we propose limiting each experiment to a maximum of six samples. Furthermore, it is advisable to conduct the entire experiment under low-temperature conditions whenever feasible.

3.1 Material Preparation

1. Harvest the strawberry fruit and promptly subject it to flash freezing using liquid nitrogen. Subsequently, store the samples at −80 °C until they are ready for utilization.
2. Prechill the cryogenic tissue grinder using liquid nitrogen. Once adequately chilled, transfer the strawberry fruit from the liquid nitrogen into the grinder, and pulverize until it reaches a fine powdered consistency.
3. Under the protection of liquid nitrogen, aliquot the ground strawberry fruit powder into 2 mL centrifuge tubes. Subsequently, freeze the samples using liquid nitrogen. Transfer the powdered fruit, along with any remaining liquid nitrogen, to a 50 mL centrifuge tube and allow liquid nitrogen to evaporate. Once evaporated, the ground tissue can be used immediately or stored at −80 °C for up to 2–3 days. It is advisable to grind the material a day prior to proceeding with the steps involved in protein extraction.

3.2 Extraction of Proteins from Strawberry Fruit

1. Prepare fresh protein extraction buffer before starting the experiment (*see* **Notes 4** and **5**).
2. Weigh 0.2 g of sample powder per tube, add 400 μL of protein extraction buffer (*see* **Note 6**), and vortex vigorously until completely thawed (attention should be paid to the ratio between fruit powder and extraction buffer at this step, ensuring optimal vortexing of the powder).
3. Incubate on ice for 30 min (*see* **Notes 7** and **8**).
4. Sonicate the samples using a cycle of 2 s on and 3 s off, totaling 20 s with an amplitude set at 40%, while maintaining low-temperature conditions during sonication (*see* **Note 9**).
5. Centrifuge at a temperature of 4 °C and a speed of 15,000 × g for a period lasting 15 min; subsequently transfer the supernatant to a new 1.5 mL centrifuge tube (*see* **Note 10**).
6. Optional: Extract a small aliquot (50 μL) from the protein extract for subsequent analysis via Western blotting.

3.3 Protein Immunoprecipitation

1. Incubate total protein extracts (approximately 500 μL) with the corresponding antibody (diluted at 1:200) for 1–2 h at 4 °C (*see* **Note 11**).
2. Add protein A + G agarose beads (diluted at a ratio of 1:10) and incubate the mixture for 1–2 h at 4 °C.
3. After completion of the immunoprecipitation reaction, remove the supernatant by centrifugation at 600 × g for 3–5 min at 4 ° C.
4. Wash the agarose beads three to five times using 1 mL extraction buffer each time, centrifuging at a speed of 600 × g for 5 min at 4 °C to eliminate nonspecifically bound proteins.
5. Ensure inclusion of a blank control during this step and elute the proteins with SDS loading buffer. The blank buffer is used to distinguish between experimental and control groups.

3.4 SDS-PAGE

1. Gel preparation: Mix acrylamide, bis-acrylamide, Tris-HCl buffer, and SDS to prepare the separating gel. Pour the mixture into a casting tray and insert a comb. Stand for 30–60 min to allow polymerization. Subsequently, prepare and pour the stacking gel on top.
2. Sample preparation: Prepare protein samples by mixing them with SDS loading buffer and boiling for 5 min to denature the proteins.
3. Loading samples: Remove the comb from the stacking gel and place the gel in an electrophoresis chamber filled with running buffer. Load the prepared protein samples into the wells.

4. Gel electrophoresis: Connect a power supply and apply a constant voltage (80–120 V) until the tracking dye reaches the bottom of the gel.
5. Gel staining: Remove the gel from the chamber and incubate in staining solution (such as Coomassie Brilliant Blue) for approximately 1 h.
6. Destaining: Destain the gel with destaining solution to enhance the visibility of protein bands.
7. Analysis: The resulting gel image can be documented using an imaging system, followed by analysis of protein bands in terms of their size and abundance.
8. After performing SDS-PAGE analysis, LC-MS/MS (nanoLC-LTQ-Orbitrap XL; ThermoFinnigan, San Jose, CA) technique can be employed for further characterization (*see* **Notes 12**, **13**, and **14**).

4 Notes

1. Tagged protein design: Utilizing epitope tags (e.g., HA, FLAG) fused to the protein of interest often improves immunoprecipitation efficiency. Ensure transgenic lines express the tagged protein in a loss-of-function mutant background (e.g., *AP1:HA* in *ap1* mutants). If complementation fails, consider altering linker sequences or testing dual-tag systems (e.g., N-terminal HA and C-terminal FLAG).
2. Antibody validation: During antibody screening, monoclonal antibodies are preferentially selected to minimize the risk of nonspecific cross-reactions, and preexperiments are conducted to verify antibody specificity and working concentration range (1:100–1:1000). Prior to IP-MS, validate antibody specificity via knockout/knockdown controls. For commercial antibodies, cross-check batch consistency using overexpression lysates. For plant endogenous protein detection, we recommend using specialized antibodies from CST (Cell Signaling Technology) or MBL (Medical&Biological Laboratories) brands.
3. Protease inhibitor preparation: Pre-dissolve protease inhibitor tablets in a small volume of buffer (e.g., 500 μL PBS) using vortexing, and then dilute into the final protein extraction buffer. Avoid prolonged storage of dissolved inhibitors (<24 h).
4. For sample preparation, we have improved the protein extraction buffer formulation by adding 1% PVPP to reduce polyphenol interference.

5. Protein extraction buffer optimization: Adjust detergent types (e.g., 0.5% IGEPAL, 0.2% SDS) based on protein solubility. For membrane-bound targets, include 0.5% sodium deoxycholate. Preclear lysates by centrifugation (15,000 × g, 10 min) to reduce debris.
6. Tissue quantity and homogenization: For recalcitrant tissues (e.g., roots, woody stems), increase homogenization time or use ceramic beads. Scale protocols for 0.1–5 g tissue, adjusting buffer volumes proportionally (e.g., 400 μL buffer per 0.2 g tissue).
7. Nuclear enrichment: If the target protein is localized in the nucleus and has low expression levels, nuclear protein enrichment and immunoprecipitation (IP) may be necessary to increase the abundance of the target protein.
8. Control experiments: Include a *GFP-only* transgenic line under the same promoter as the bait protein. For label-free quantification, spike in a non-plant protein (e.g., BSA) as an internal standard. Three negative control experiments are also needed: (1) negative control with empty beads, (2) negative control with isotype IgG, and (3) negative control with mutant.
9. Sonication parameters: Optimize amplitude (20–40%) and pulse duration (up to 5-s on/10-s off) to prevent overheating. For dense tissues, perform sonication in a chilled ethanol bath.
10. Column clogging prevention: Prefilter lysates through 0.45 μm PVDF membranes before loading onto IP columns. For magnetic beads, use low-binding tubes to minimize bead loss.
11. Quality control checkpoints: Retain aliquots pre-IP, post-wash, and post-elution for SDS-PAGE and Coomassie staining. Compare band intensities to assess IP efficiency and nonspecific binding.
12. Protease selection for digestion: For arginine-rich proteins, use Lys-C instead of trypsin. Validate digestion efficiency via peptide fingerprinting.
13. Posttranslational modification (PTM) analysis: Enrich phosphorylated peptides using TiO2 beads or IMAC prior to MS. Note that PTM analysis may require extended gradient times.
14. When analyzing data, we recommend jointly searching the TAIR and UniProt plant protein databases while incorporating common contaminant databases to exclude interferences. Meanwhile, triple negative controls are set up, and PRM technology is used for targeted validation.

Conflict of Interest Statement

The authors have no conflicts of interest to declare that are relevant to the content of this chapter.

References

1. Gromiha MM, Yugandhar K, Jemimah S (2017) Protein-protein interactions: scoring schemes and binding affinity. Curr Opin Struct Biol 44: 31–38. https://doi.org/10.1016/j.sbi.2016.10.016
2. Jones S, Thornton JM (1996) Principles of protein-protein interactions. Proc Natl Acad Sci USA 93:13–20. https://doi.org/10.1073/pnas.93.1.13
3. Slater O, Miller B, Kontoyianni M (2020) Decoding protein-protein interactions: an overview. Curr Top Med Chem 20:855–882. https://doi.org/10.2174/1568026620666200226105312
4. Aebersold R, Mann M (2003) Mass spectrometry-based proteomics. Nature 422: 198–207. https://doi.org/10.1038/nature01511
5. Witze ES, Old WM, Resing KA, Ahn NG (2007) Mapping protein post-translational modifications with mass spectrometry. Nat Methods 4:798–806. https://doi.org/10.1038/nmeth1100
6. Johnson ES (2004) Protein modification by SUMO. Annu Rev Biochem 73:355–382. https://doi.org/10.1146/annurev.biochem.73.011303.074118
7. Kwon SJ, Choi EY, Choi YJ et al (2006) Proteomics studies of post-translational modifications in plants. J Exp Bot 57:1547–1551. https://doi.org/10.1093/jxb/erj137
8. Swatek KN, Komander D (2016) Ubiquitin modifications. Cell Res 26:399–422. https://doi.org/10.1038/cr.2016.39
9. Ubersax JA, Ferrell JEJ (2007) Mechanisms of specificity in protein phosphorylation. Nat Rev Mol Cell Biol 8:530–541. https://doi.org/10.1038/nrm2203

Chapter 10

TurboID-Based Proximity Labeling in Plants

Chaonan Shi, Huang Tan, and Rosa Lozano-Durán

Abstract

Proximity labeling (PL) techniques have emerged as powerful approaches for the study of protein associations in vivo. In this chapter, we present a step-by-step TurboID-based PL protocol to define the "proxiome" of a protein of interest in plants, from the generation of the fusion protein to the analysis of the results, using the publicly available Ti-TAN plasmid collection.

Key words TurboID, Proximity labeling, Proximal proteome, Proxiome, Protein complexes, Protein-protein interactions, *Nicotiana benthamiana*

1 Introduction

Proteins are the building blocks of life, and their interactions and complex formation play a central role in virtually all biological processes. Therefore, the investigation of protein-protein interactions is a fundamental stepping stone in life sciences, and multiple experimental techniques have been established for this purpose. However, while some interactions are strong and stable, others are weak and/or transient, and their detection in vivo presents additional challenges.

The past decades have witnessed the development of an increasing number of methods to detect protein-protein interactions in plants. Some of these techniques can be used as unbiased approaches to screen for interactors of a protein of choice, employed as bait, against the whole plant proteome; a widely used example of this kind is affinity purification followed by mass spectrometry (AP-MS). Other methods, such as Förster resonance energy transfer by fluorescence lifetime imaging (FRET-FLIM), split-luciferase, bimolecular fluorescence complementation (BiFC), or co-immunoprecipitation, serve to test the interaction between two known proteins of interest [1]. Unfortunately, these

José M. Franco-Zorrilla and Carmen Martín-Pizarro (eds.), *Plant Transcription Factors: Methods and Protocols*, Methods in Molecular Biology, vol. 2985, https://doi.org/10.1007/978-1-0716-4972-5_10,

experiments cannot, in most cases, capture short-lived or weak interactions, which therefore remain difficult to identify.

Recently, biotin ligase-based proximity labeling (PL) approaches have joined the portfolio of protein-protein interaction techniques as a powerful in vivo alternative to the aforementioned methods. PL techniques not only can detect dynamic interactions, but they do so in the natural physiological and cellular context [2]. In PL assays, a genetically encoded biotin ligase enzyme, engineered to confer promiscuity, is translationally fused to the protein of interest (POI), enabling biotinylation of proteins within a 10-nm radius of the fusion protein when biotin is provided. Once the enzymatic reaction has taken place, biotinylated (proximal) proteins can be affinity-purified using streptavidin beads and identified by mass spectrometry analyses.

In plants, different engineered biotin ligases have been tested so far, including BioID (BirA*) [3], TurboID [4, 5], and AirID [6]. Based on the data available, TurboID is considered to offer superior performance, since it can efficiently label proximal proteins in a matter of minutes at a wide range of temperatures [2]. Recently, with the aim to contribute to the accessibility of this technique to plant scientists worldwide, we generated an easy-to-use set of vectors for TurboID assays in plants called *TurboID Transient Assays in Nicotiana benthamiana* (Ti-TAN) plasmid toolbox, available to the scientific community [7].

It is uncontested that TurboID-based PL entails great potential for the study of protein-protein interactions, protein networks, and subcellular proteomes; however, the use of appropriate controls is essential to guarantee reliability of the results. More specifically, positional effects of the TurboID fusion both on subcellular localization and enzymatic activity have to be tested beforehand, and suitable fusion proteins must be used as negative controls in the identification of proximal proteomes. In this chapter, we will guide the reader through the generation of fusion proteins for TurboID-based PL using the Ti-TAN vectors and explain how to design and perform the necessary controls for the subsequent proximity labeling assays. We also describe how to carry out the discovery proximity labeling assays and the ensuing analysis. To help with orientation, the workflow of the procedure is shown in Fig. 1.

2 Materials

2.1 Cloning

1. Primers: The Ti-TAN plasmid collection contains Gateway-compatible and Golden Gate-compatible vectors (*see* **Note 1**). For the Gateway cloning system, primer design should be as follows—forward primer: GGGG-attB1 (GGGGA CAAGTTTGTACAAAAAAGCAGGCTCC)-template-specific sequence; reverse primer: GGGG-attB2 (

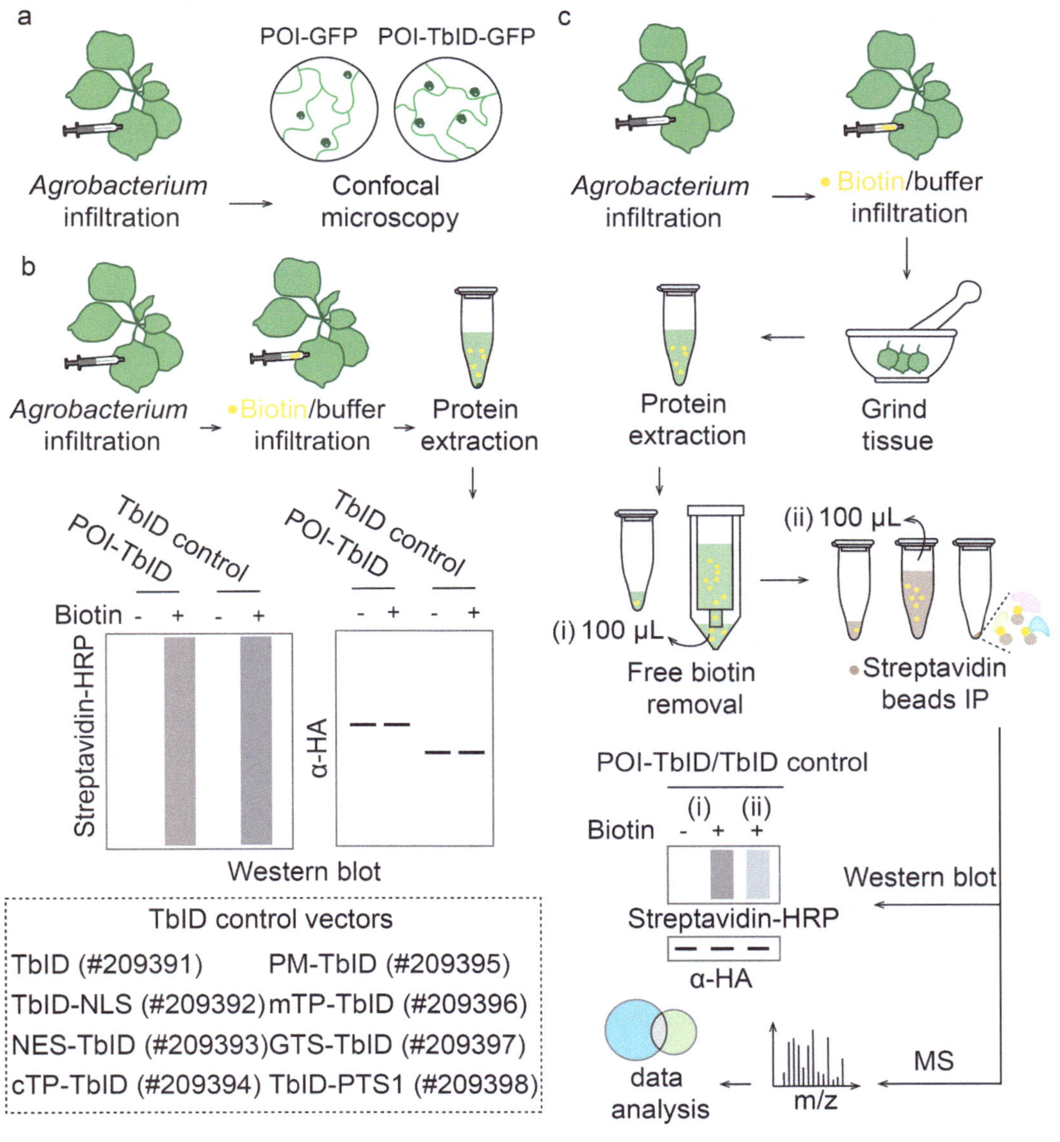

Fig. 1 Schematic workflow of the TurboID-based proximity labeling (PL) assay. (**a**) Observation of the subcellular localization of a GFP-tagged protein of interest (POI), either with or without TurboID (TbID). This step aims to confirm that the subcellular localization of the POI remains unaffected by the TbID fusion. Related to Subheading 3.3. (**b**) Evaluation of the protein expression and the biotinylation activity of the TbID-fused POI. A control TbID, displaying the same subcellular localization pattern as the TurboID-fused POI, should be utilized as a negative control. The biotinylated signal is detected by streptavidin-HRP; an α-HA antibody is used to examine the accumulation of fusion proteins. HRP: horseradish peroxidase. The TbID control vectors available in the Ti-TAN plasmid toolbox and the corresponding Addgene catalog numbers are shown. Related to Subheading 3.4. (**c**) Detection of the proximal proteins of POI via mass spectrometry (MS) analysis. A desalting column is used to remove the excess biotin in the lysate. Biotinylated proteins are enriched by streptavidin-coated magnetic beads. The resulting proteins are submitted to MS analysis for identification. Related to Subheading 3.5

GGGGACCACTTTGTACAAGAAAGCTGGGTC)-template-specific sequence (*see* **Note 2**). For the Golden Gate ready-to-use plasmids, primer design should be as follows—forward primer: TCCGTCTCTCACC-template-specific sequence; reverse primer: ATCGTCTCACCTT-template-specific sequence (*see* **Notes 2** and **3**).

2. Enzymes: Q5® High-Fidelity DNA Polymerase for gene amplification; for Gateway cloning, Gateway™ BP Clonase™ II Enzyme mix and Gateway™ LR Clonase™ II Enzyme mix; for Golden Gate cloning, Esp3I restriction enzymes.
3. Ti-TAN destination vectors [7] (*see* **Note 1**).
4. Luria-Bertani broth (LB) medium: Add 10 g NaCl, 10 g tryptone, and 5 g yeast extract and dissolve them in 900 mL distilled water (dH_2O). Adjust the final volume to 1000 mL for liquid LB. 15 g of agar are added to 1 L for solid LB. Autoclave at 121 °C for 15 min. Cool down to 60 °C and add appropriate antibiotics. Pour into plates for solid LB. Store plates and/or liquid LB at 4 °C.
5. Antibiotics: 50 mg/mL kanamycin is dissolved in dH_2O; 50 mg/mL spectinomycin is dissolved in dH_2O; 50 mg/mL rifampicin is dissolved in dimethylsulfoxide (DMSO); 50 mg/mL gentamycin is dissolved in dH_2O. Antibiotic solutions are sterilized by filtration through 0.22 μm membrane filters. Store antibiotics at −20 °C until use.
6. Competent cells: *Escherichia coli* strain TOP10 (*see* **Note 4**) and DB3.1 (for propagating plasmids containing the *ccdB* gene); *Agrobacterium tumefaciens* strain GV3101 (*see* **Note 4**).

2.2 Agroinfiltration and Biotin Treatment

1. Infiltration buffer: 10 mM $MgCl_2$, 10 mM MES – KOH (adjust pH to 5.2), and 150 μM acetosyringone. Infiltration buffer should be prepared each time freshly.
2. Biotin solution: 50 μM biotin and 2.5 mM $MgCl_2$ (*see* **Note 5**).
3. pH meter.
4. Syringes, needles.
5. Spectrophotometer.
6. 4-week-old *Nicotiana benthamiana* plants (for transient expression assays).

2.3 Subcellular Localization

1. Cork borers.
2. Microscopic slides.
3. Cover slips.
4. Tweezers.

2.4 Protein Extraction and Western Blot

1. Liquid nitrogen.
2. Aluminum foil.
3. Mortars and pestles.
4. Protein extraction buffer: 100 mM Tris-HCl, pH 7.5, 150 mM NaCl, 10% glycerol, 5 mM EDTA, 5 mM DTT, 1% protease inhibitors cocktail, 1 mM PMSF, 1% NP40.
5. 6× loading buffer: 50 mM Tris-HCl, pH 6.8, 5% glycerol, 1% SDS, 0.017% bromophenol blue (*see* **Note 6**).
6. Antibodies: Anti-HA antibody (e.g., Sigma, cat. 11583816001, diluted in 5% milk in 1:1000) and streptavidin-HRP (e.g., Abcam, cat. ab7403, diluted in 3% BSA in 1:20000) (*see* **Note 7**).

2.5 Preparation of Samples for MS Analysis

1. PD-10 desalting columns.
2. Equilibration buffer: 100 mM Tris-HCl, pH 7.5, 150 mM NaCl, 10% glycerol, 5 mM EDTA, 2 mM DTT.
3. Streptavidin-C1-conjugated magnetic beads.
4. Wash buffer I: 0.5% SDS.
5. Wash buffer II: 50 mM HEPES, pH 7.5, 500 mM NaCl, 1 mM EDTA, 1% Triton-X-100, 0.5% Na-deoxycholate.
6. Wash buffer III: 10 mM Tris-HCl, pH 7.5, 250 mM LiCl, 1 mM EDTA, 0.5% NP-40, 0.5% Na-deoxycholate.
7. 50 mM Tris-HCl, pH 7.5.
8. 1× PBS solution.

2.6 General Equipment

1. Incubators (37 °C, 28 °C).
2. Adjustable shakers (37 °C, 28 °C).
3. Centrifuges.
4. TissueLyser II.
5. Rotator.
6. Confocal microscopy.

3 Methods

3.1 Cloning

This protocol is designed for cloning in vectors of the Ti-TAN plasmid toolbox [7]; however, after the cloning step, the protocol is applicable to any TurboID fusion, transiently or stably expressed from any binary vector with the appropriate selection markers. The cloning procedure is illustrated in Fig. 2.

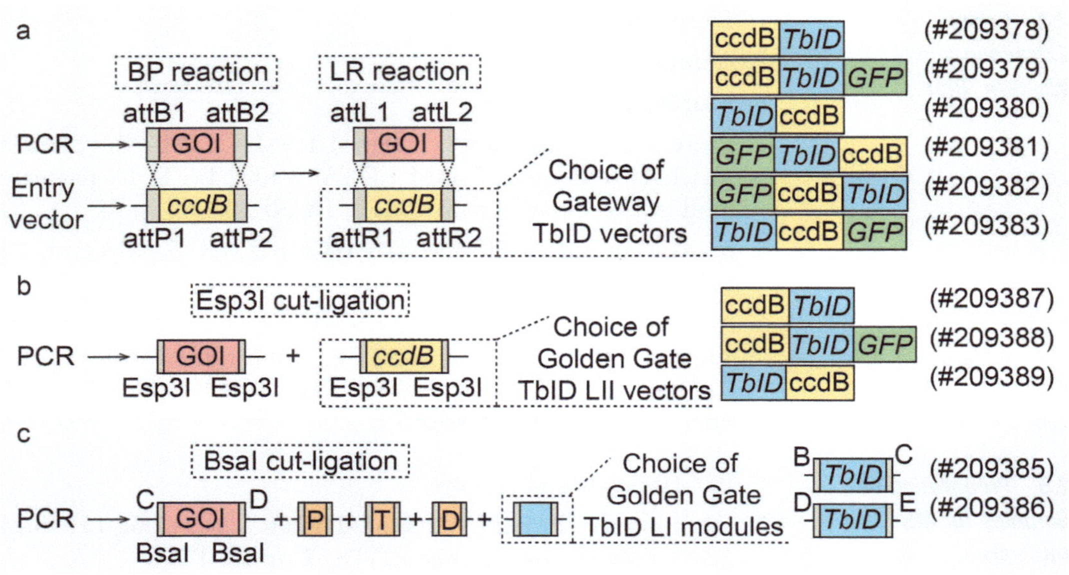

Fig. 2 A simplified cloning strategy for TurboID-based proximity labeling (PL) constructs by using the Ti-TAN plasmid toolbox. Gateway (**a**) and Golden Gate cloning options (**b, c**) are shown. The TbID vectors in the Ti-TAN toolbox and the corresponding Addgene catalog numbers are indicated on the right

3.1.1 Gateway Cloning

1. PCR-amplify the coding sequence of the gene of interest (GOI) using gene-specific primers for the Gateway system (see Subheading 2.1) and purify the PCR products.
2. Clone the GOI into a Gateway entry vector (e.g., pDONR™/Zeo, Thermo Scientific) through a BP reaction following the manufacturer's instructions.
3. Transform the BP reaction into *E. coli* competent cells following your preferred transformation protocol. Grow them on selective LB plates containing appropriate antibiotics and incubate them at 37 °C overnight.
4. Confirm transformants by colony PCR, culture a positive colony, extract plasmid, and verify the cloned sequence by Sanger sequencing.
5. Subclone the GOI into the appropriate Gateway-compatible TurboID vectors with or without a green fluorescent protein (GFP) tag (*see* **Notes 8** and **9**) via an LR reaction following the manufacturer's instructions. Proceed as indicated in **step 4**.

3.1.2 Golden Gate Cloning

1. PCR-amplify the coding sequence of GOI using gene-specific primers for the Golden Gate system (see Subheading 2.1) and purify the PCR products (*see* **Notes 10** and **1**).
2. Insert the PCR fragments into the ready-to-use level II TurboID vectors (50–100 ng/μL) via Esp3I cut-ligation protocol.

The insert is mixed with a vector in a molar ratio of 1:3 to 1:10. For a 10 μL reaction, use 0.8 μL Esp3I, 1 μL T4 ligase, and 1 μL T4 ligase buffer.

3. Incubate the reaction in a thermocycler for 30–50 cycles, cycling between 37 °C and 16 °C for 5 min each, followed by 50 °C for 5 min and 80 °C for 5 min.
4. Transform 3–5 μL of the reaction into *E. coli* competent cells. Store remaining ligation at −20 °C. The subsequent steps follow the Gateway cloning section (**steps 3** and **4**).

3.2 Agroinfiltration and Biotin Treatment

1. Transform expression constructs generated in Subheading 3.1 into *A. tumefaciens* strain GV3101 competent cells via either electroporation or freeze/thaw method. Plate them onto solid LB plates containing appropriate antibiotics, and incubate at 28 °C for 2 days.
2. Pick single colonies and streak them onto new solid LB plates with appropriate antibiotics. Subsequently, conduct colony PCR to verify the presence of the GOI.
3. Inoculate liquid cultures from the positive colonies harboring the desired construct in liquid LB with appropriate antibiotics overnight at 28 °C.
4. Collect the pellet by centrifugation at 3000 × g for 10 min at room temperature (RT), resuspend the pellet in the infiltration buffer, and adjust the optical density (OD_{600}) to 0.2–0.5 (*see* **Note 11**).
5. Keep the *A. tumefaciens* cultures in the dark for 2–4 h (*see* **Note 12**).
6. Infiltrate fully expanded leaves from 4- to 5-week-old *N. benthamiana* plants with a 1 mL needleless syringe (*see* **Note 13**).
7. Leave the plants in the growth room for 48 h (*see* **Note 14**). Proceed directly to Subheading 3.3 if the samples are intended for observation of subcellular localization.

3.3 Observation of the Subcellular Localization of TurboID-Fused Proteins

1. Collect a leaf disc from leaves expressing a TurboID-fused protein of interest (POI) tagged with GFP or GFP-fused POI, place it on a microscopic slide, add a drop of water, and cover with a glass coverslip.
2. Image under confocal microscopy, following the manufacturer's instructions. The preset settings for GFP are excitation at 488 nm and emission at 500–550 nm. Observe the subcellular localization of the TurboID-fused POI tagged with GFP and GFP-fused POI. Marker proteins can be used if desired.
3. Determine which negative control(s) should be used for the experiment based on the subcellular localization observed from

step 2 (*see* **Note 15**). An appropriate negative control must display a subcellular distribution similar to that of the POI. The TurboID-fused POI (without GFP tag) and appropriate negative controls will be used in Subheading 3.4.

3.4 Detection of Protein Accumulation and Biotin Ligase Activity (See Note 16)

1. Following the agroinfiltration procedure in Subheading 3.2, infiltrate *Agrobacterium* cultures harboring TurboID-fused POI or control constructs into *N. benthamiana* fully expanded leaves, respectively.
2. At 42 h post-infiltration (hpi) (*see* **Note 17**), infiltrate biotin buffer or mock solution (2.5 mM $MgCl_2$) onto the leaves expressing the POI or negative controls. An ideal negative control is the TurboID displaying the same subcellular localization than the POI (based on the results obtained in Subheading 3.3; *see* **Note 18**).
3. Harvest samples 6 h after biotin treatment (*see* **Note 19**) and proceed to **step 4** or store the samples in the −80 °C freezer.
4. Grind six to ten 6 mm leaf discs from biotin- and $MgCl_2$-treated samples into a fine powder with a TissueLyser II (*see* **Note 20**) in a 2 mL tube.
5. Add protein extraction buffer (2 mL/g tissue) to the tubes.
6. Vortex samples thoroughly and incubate them on a rotator at 4 °C for 15–30 min.
7. Centrifuge at 16,500 × g at 4 °C for 15 min.
8. Transfer the supernatant to a new 2 mL tube and add 6× loading buffer containing 50 mM DTT to each tube.
9. Denature the samples by boiling the mixture at 70 °C for 20 min.
10. Centrifuge at 12,000 × g for 1 min, and take the supernatant for Western blot analysis using the corresponding antibodies to detect protein accumulation and biotin ligase activity (*see* **Note 21**) (with and without addition of exogenous biotin).
11. Analyze the immunoblotting results. If biotinylated proteins are enriched in the sample expressing TurboID-fused POI compared to control sample not expressing any TurboID, and protein accumulation of the TurboID-fused POI is detected, the next section can be performed (*see* **Note 22**).

3.5 Sample Preparation for MS

1. Infiltrate *Agrobacterium* cells carrying the constructs to express the TurboID-fused constructs or control vectors into leaves of 4- to 5-week-old *N. benthamiana* plants (*see* **Note 23**). To proceed with biotin infiltration, follow Subheading 3.4, **steps 2** and **3**.

2. Collect the entire infiltrated leaves and remove the leaf middle vein. Store the leaf samples at −80 °C or proceed to grind the tissue to a fine powder using a pestle and mortar.
3. Repeat Subheading 3.4, **steps 5–7**. If 1 g of ground tissue (*see* **Note 23**) is used, transfer about 0.5 g of leaf powder to a 2 mL tube and add 1 mL protein extraction buffer (two tubes for each sample). Take 100 μL of protein extract to serve as an input for Western blot analysis.
4. Equilibrate the PD-10 desalting column using equilibration buffer, following the manufacturer's manual (both gravity and spin protocols are provided; we use the gravity protocol as an example here).
5. To remove the free biotin, the protein extract from **step 3** is added to the center of the equilibrated desalting column. Put the columns in a 50 mL Falcon tube and centrifuge at 1000 × g at 4 °C for 2 min. The flow-through is transferred to a new 2 mL tube and kept on ice. About 100 μL of desalted protein extract is taken to serve as an input for Western blot.
6. Wash the streptavidin-C1-conjugated magnetic beads twice in the protein extraction buffer at RT. For each sample, 80 μL of beads are used (*see* **Note 24**).
7. Transfer the desalted protein extracts to the prewashed beads, and incubate at 4 °C for 3–4 h on a rotator to enrich the biotinylated proteins (*see* **Note 25**).
8. Capture the beads using a magnetic rack for 2 min at RT. Gently remove the supernatant by pipetting.
9. Add 1 mL wash buffer I to the tube and keep it on a rotator at RT (*see* **Note 26**) for 5–10 min. The beads are captured as above.
10. Add 1 mL wash buffer II to the tube and keep it on a rotator at 4 °C for 5–10 min. The beads are captured as above.
11. Add 1 mL wash buffer III to the tube and keep it on a rotator at 4 °C for 5–10 min. The beads are captured as above.
12. Add 1 mL 50 mM Tris-HCl, pH 7.5, to the tube to remove the detergent and keep it on a rotator at 4 °C for 5–10 min. Repeat once. The beads are captured as above.
13. Wash the beads in 1 mL 1× PBS solution three times on a rotator at 4 °C for 5–10 min each. After the wash, take 100 μL of the mixture for Western blot analysis to confirm the enrichment of biotinylated proteins (*see* **Note 27**).
14. Put the rest of the mixture back on the magnetic rack for 2 min and remove all supernatant. The dry beads can be flash-frozen in liquid nitrogen and stored at −80 °C or sent immediately for LC-MS/MS analysis.

3.6 Data Analysis

The data from mass spectrometry can be visualized using the Scaffold software. Some selection criteria need to be chosen. As an example, for most applications, we use the following:

1. Proteins must have protein and peptide thresholds higher or equal to 95%.
2. Proteins are present in at least two out of three replicates.
3. The peptide numbers in the experimental sample are at least twice as those in the negative control.

4 Notes

1. The GOI and TurboID sequence can be fused by other methods, e.g., overlapping PCR; choose the appropriate destination vectors accordingly.
2. CTA, TCA, and TTA are the reverse complementary stop codons. Add or remove according to the tag position in the destination vector.
3. The GOI should not contain Esp3I cutting sites.
4. Other alternative strains can be used.
5. The concentration of biotin should be adjusted depending on the POI and the purpose of the experiments. This should be determined experimentally.
6. 50 mM DTT must be added to the 6x loading buffer before using.
7. The antibody dilution will depend on the brand; please follow the manufacturer's instructions.
8. The TurboID can be fused at either the N- or the C-terminus of the POI, as long as the fusion does not affect subcellular localization (check via confocal microscopy; see Subheading 3.3), protein accumulation (check via Western blot; see Subheading 3.4), and function, if this can be assessed by complementation.
9. The GFP-containing TurboID vectors should be used to confirm that the TurboID fusion has no impact on the subcellular localization of the POI.
10. Alternatively, instead of using the ready-to-use LII vectors, you can generate the constructs expressing TurboID-fused POI by combining appropriate LI modules and dummies. In the Ti-TAN toolbox, the TurboID is placed at LI B-C or D-E module; therefore, the GOI should be cloned in LI C-D module.
11. For co-infiltration of two or three constructs, mix them as 1:1 or 1:1:1 ratio by volume.

12. Incubation of the *Agrobacterium* culture in the dark before infiltration helps to induce the *Agrobacterium vir* genes by acetosyringone, which further promotes efficient transformation.
13. Depending on the needs, one completely infiltrated leaf is enough to test if the desired proteins accumulate well and have biotinylation activity; four or more leaves are needed for MS analysis.
14. The time between infiltration and maximum accumulation of the POI needs to be experimentally determined (and usually ranges between 1 and 3 days).
15. This section is critical to make sure that the fusion of TurboID has no impact on the localization of a POI.
16. To this purpose, a small amount of sample, e.g., 0.5 g, is enough for further analysis; this section aims to test if the TurboID-fused proteins accumulate and maintain biotin ligase activity.
17. It is recommended to experimentally determine the time when the optimal accumulation of the POI is reached before biotin infiltration.
18. The Ti-TAN plasmid toolbox contains TurboID controls in various subcellular localizations, including cytosol, nucleus, mitochondria, chloroplasts, Golgi apparatus, and peroxisomes [7].
19. We routinely use a 6-hour incubation with biotin, following [5]. It is advisable to experimentally optimize the incubation time when working with different POIs and other plant species.
20. Alternative tools, e.g., micropestles, can be used.
21. An anti-HA antibody is used to detect protein accumulation (the TurboID fragment contains 3X HA tags); streptavidin-HRP antibody is used to detect biotinylated proteins.
22. Different plants and tissues might have different endogenous levels of biotin that would lead to no observable difference between—and + biotin.
23. The amount of sample required is determined by the accumulation of your POI.
24. The volume of beads used for IP depends on the amount of the sample, e.g., 80 μL is sufficient for 2 g sample.
25. Longer incubation times can be used. However, longer incubations might increase the risk of protein degradation and nonspecific protein binding.
26. This wash step must be carried out at RT to prevent SDS from precipitating at temperatures below 16 °C; subsequent wash

steps (Subheading 3.5 **steps 8–11**) should be conducted at 4 °C to minimize the risk of protein degradation.

27. Western blot can also be used to test the proximity of two proteins of choice (POI-A and POI-B). To this end, co-infiltrate constructs carrying TurboID-fused POI-A and POI-B fused with a tag (e.g., GFP) onto *N. benthamiana* leaves, perform streptavidin immunoprecipitation, and detect the POI with an appropriate antibody.

Acknowledgments

Work in RLD's lab is partially funded by the Excellence Strategy of the German Federal and State Governments.

Conflict of Interest The authors have no conflicts of interest to declare that are relevant to the content of this chapter.

References

1. Strotmann VI, Stahl Y (2022) Visualization of in vivo protein-protein interactions in plants. J Exp Bot 73:3866–3880. https://doi.org/10.1093/jxb/erac139
2. Xu S-L, Shrestha R, Karunadasa SS, Xie P-Q (2023) Proximity labeling in plants. Annu Rev Plant Biol 74:285–312. https://doi.org/10.1146/annurev-arplant-070522-052132
3. Kim DI, Birendra KC, Zhu W et al (2014) Probing nuclear pore complex architecture with proximity-dependent biotinylation. Proc Natl Acad Sci USA 111:E2453–E2461. https://doi.org/10.1073/pnas.1406459111
4. Branon TC, Bosch JA, Sanchez AD et al (2018) Efficient proximity labeling in living cells and organisms with TurboID. Nat Biotechnol 36: 880–887. https://doi.org/10.1038/nbt.4201
5. Zhang Y, Song G, Lal NK et al (2019) TurboID-based proximity labeling reveals that UBR7 is a regulator of N NLR immune receptor-mediated immunity. Nat Commun 10:3252. https://doi.org/10.1038/s41467-019-11202-z
6. Kido K, Yamanaka S, Nakano S et al (2020) AirID, a novel proximity biotinylation enzyme, for analysis of protein-protein interactions. elife 9. https://doi.org/10.7554/eLife.54983
7. Tan H, Zhou Y, Dinius E, Lozano-Durán R (2024) The Ti-TAN plasmid toolbox for TurboID-based proximity labeling assays in Nicotiana benthamiana. J Integr Plant Biol 66: 166–168. https://doi.org/10.1111/jipb.13610

Chapter 11

FRET-FLIM: Unveiling Transcription Factor Interactions in Plants

María Florencia Perotti, Carlos Sánchez-Gómez, Carmen Martin-Pizarro, and David Posé

Abstract

Gene regulatory proteins, such as transcription factors (TFs), bind to DNA and orchestrate spatial and temporal gene expression patterns. This regulation relies on dynamic and coordinated protein-protein interactions, either through direct pairwise interactions or as part of multiprotein complexes. Therefore, the development of methods to assay these interactions within the cellular context is crucial. Förster resonance energy transfer combined with fluorescence lifetime imaging microscopy (FRET-FLIM) is a powerful quantitative imaging technique for detecting protein-protein interactions in vivo. This approach involves labeling the proteins of interest with fluorescent tags and measuring changes in the fluorescence lifetime of the donor fluorophore. The reduction in donor lifetime in the presence of the acceptor fluorophore provides direct evidence of a physical interaction between the proteins under study. In this chapter, we present a detailed and simple protocol for the acquisition and analysis of FRET-FLIM data using the Leica STELLARIS 8 FALCON FLIM Microscope® system, illustrated with two interacting transcription factor proteins as an example.

Key words FRET-FLIM, Protein-protein interactions, Confocal microscopy, Transcription factors, Fluorophores pairs, In planta, In vivo

1 Introduction

The growth pattern of a plant is shaped by its remarkable ability to perceive environmental cues and exhibit developmental plasticity. Upon receiving external signals, plants initiate a series of signaling cascades that lead to cellular responses involving changes in gene expression, metabolism, or protein degradation. Consequently, these signaling pathways are tightly regulated to ensure appropriate physiological outcomes. At the transcriptional level, transcription factors (TFs) play a central and dynamic role as key regulators of gene expression in plants. TFs function by binding DNA as homo- or hetero-oligomers and by directly interacting with components of the basal transcriptional machinery, coactivators, or

José M. Franco-Zorrilla and Carmen Martín-Pizarro (eds.), *Plant Transcription Factors: Methods and Protocols*, Methods in Molecular Biology, vol. 2985, https://doi.org/10.1007/978-1-0716-4972-5_11,

chromatin-modifying factors, often forming large multiprotein assemblies. Moreover, some of these interactions remain stable, while others are transient depending on the biological context [1]. To study such dynamic protein-protein interactions, several in vitro and in vivo techniques have been developed [2–4]. Among the most widely used approaches are co-immunoprecipitation (Co-IP) or yeast two-hybrid (Y2H) assays. These methods are highly effective for identifying potential interaction candidates during the initial stages of research [5, 6] and are commonly accepted as standard procedures for validating protein-protein interactions [7, 8]. However, these techniques have intrinsic limitations, like the lack of spatial and temporal resolution [3]. In contrast, techniques such as bimolecular fluorescence complementation (BiFC), split-luciferase (splitLuc), and Förster resonance energy transfer (FRET)-based, among other in vivo assays, enable the detection of protein-protein interactions within living cells, providing spatiotemporal insight into interaction dynamics. These techniques rely on the expression of proteins fused to fluorescent or luminescent tag proteins that emit detectable signals within specific spectral ranges [2].

At the molecular level, BiFC and splitLuc are based on the reconstitution of a reporter signal [9]. In BiFC, two nonfluorescent fragments of a fluorescent protein (e.g., GFP, YFP, RFP) are fused to the proteins of interest; if the proteins interact, the fluorescent protein is reconstituted, resulting in detectable fluorescence [10]. In splitLuc assays, the luciferase enzyme is similarly split into two fragments that reassemble upon protein interaction to emit bioluminescence [11]. However, a notable drawback of BiFC is the frequent occurrence of false-positive protein-protein interactions due to the inherent affinity between the two fragments of the fluorescent protein. Hence, it is essential to include appropriate negative controls to validate results [12]. In contrast, the luciferase fragments do not spontaneously reconstitute, reducing the likelihood of false-positive interactions. Additionally, because firefly luciferase reassembly is reversible, splitLuc assays are well suited for analyzing the dynamic formation and dissociation of protein complexes [13]. However, both BiFC and splitLuc assays have significant limitations: They are not highly sensitive, do not allow for the distinction between direct and indirect interactions, and lack a quantitative assessment of interaction strength.

Alternatively, FRET-based techniques offer an alternative approach with higher sensitivity and spatiotemporal resolution [14, 15]. FRET relies on the principle that energy can be transferred between two fluorophores when a donor's emission spectrum overlaps with the absorption spectrum of an acceptor, resulting in a reduction in the fluorescence emission intensity and lifetime (τ) of the former, accompanied by an increase in the fluorescence emission intensity of the acceptor. This energy transformation only occurs when donors and acceptors are in a favorable orientation and within extremely close proximity (<10 nm), which makes it a sensitive technique. Therefore, to implement this method for protein-protein interaction analyses, proteins of interest are fused to donor and acceptor fluorescent proteins, and interaction is inferred from changes in fluorescence intensity and lifetime.

In this context, different approaches are used to detect FRET in microscopy. The first group involves the quantification of the fluorescence intensity of either the donor (acceptor-photobleaching, FRET-APB) or the acceptor (sensitized emission, FRET-SE). However, these techniques are quite susceptible to variations in the relative concentration of both fluorophores [2]. On the other hand, FRET can be quantified by the detection of the fluctuations in the fluorescence lifetime of the donor fluorophore, which is defined as the average time a fluorophore remains excited after excitation before returning to the ground state by emitting a photon. Thus, the non-radiative energy transfer of the donor to the acceptor involves a decrease in the donor's lifetime. In this context, the fluorescence lifetime imaging microscopy for measuring FRET (FRET-FLIM) enables the quantification of the decrease in the donor's lifetime when proteins of interest are fused to a specific fluorophore pair. By measuring this reduction, FRET-FLIM makes it possible to distinguish and quantify the interaction between proteins, providing insights into molecular proximity and dynamics in the biological context [13]. Moreover, the FRET-FLIM process allows precise and sensitive measurements regardless of concentration changes, as it requires only the donor fluorescence lifetime to be quantified. This combination of high sensitivity and quantitative capability sets FRET-FLIM apart as a powerful tool for studying protein-protein interactions [2]. For instance, FRET-FLIM has been used to elucidate complex protein-protein interaction systems in plants [16–20]. Here, we describe the FRET-FLIM methodology for determining protein-protein interactions between transcription factors following transient protein expression in *Nicotiana benthamiana* leaves. Specifically, we will illustrate an example using the Leica STELLARIS 8 FALCON FLIM Microscope® system and two interacting proteins from the MYB and the bHLH TF families.

2 Materials

2.1 Expression Vectors and Plants

1. Expression vectors: Choose a system that allows constitutive expression of the protein of interest fused at its N- or C-terminus to a fluorescent protein (*see* **Note 1**). For example, the following Gateway system vectors® allow N- or C-terminal fusion of the protein of interest to eGFP or mRFP: pK7WGF2

Table 1
FRET-FLIM pairs are employed frequently in plant research

Donor	Acceptor	λ Donor excitation	λ Acceptor emission
eGFP	mRFP	488	600
ECFP	EYFP	434	527
mTRQ1	mVenus	434	527
mEGFP	mCherry	488	610

(N-terminal fusion to eGFP), pK7FWG2 (C-terminal fusion to eGFP), pGWB554 (C-terminal fusion to eRFP), and pGWB555 (N-terminal fusion to mRFP). Selecting the appropriate fluorophore pairs for FRET-FLIM is essential. A common challenge in FRET-FLIM is the significant spectral overlap between the chosen fluorophore pairs. Issues arise when fluorescence from the donor molecule bleeds into the acceptor channel or when the acceptor is directly excited by the laser intended for the donor. Examples of FRET-FLIM pairs are detailed in Table 1 [13].

2. For accurate measurements, both positive and negative controls are necessary. The donor (eGFP) protein alone, without a possible interaction, is typically used as the main negative control. However, including as a negative control a protein combination known not to interact is recommended. If possible, proteins known to interact should be used as positive controls.
3. *Agrobacterium tumefaciens* transformed with protein constructs of interest.
4. Plant material: 3- to 4-week-old *Nicotiana benthamiana* plants grown under long-day conditions (16-h light/8-h dark) at 22 °C.

2.2 Microscopy Features and Software

1. The FRET-FLIM technique requires a microscope with specific components. Among these features, the following are essential: a pulsed laser for fluorescence excitation of the proteins of interest, high-quality objectives, sensitive and fast detectors, filters or a spectrometer to select specific emission wavelengths of donor and acceptor fluorophores, and a Time-Correlated Single Photon Counting (TCSPC) system for precise measurement of fluorescence lifetimes. In this example, a Leica STELLARIS 8 FALCON FLIM Microscope® is used. This microscope has a pulsed white light laser (WLL), which provides broad spectral excitation from 440 to 790 nm, allowing flexible and simultaneous use of multiple laser lines, as well as high-efficiency detectors (Power HyD X and Power HyD R) and hybrid detectors (HyD detectors) that are specifically designed for FRET-FLIM analysis.

2. Microscopy equipment for FRET-FLIM is integrated with control and analysis software for data acquisition and analysis of FRET efficiency and fluorescence lifetimes. All experimental data described in this chapter are visualized and analyzed using the Leica LAS X FLIM FCS®.
3. Microscope slides and coverslips.

2.3 Buffers and Medium Composition

1. Infiltration buffer: 10 mM $MgCl_2$, 10 mM 2-(*N*-morpholino) ethanesulfonic acid (MES), pH 5.6, 100 μM acetosyringone.
2. Luria-Bertani medium (LB): 1% Bacto-tryptone, 0.5% yeast extract, 1% NaCl.

3 Methods

The FRET-FLIM assay involves the transient expression of proteins (in this case, TFs) in *N. benthamiana* leaves, followed by FRET-FLIM image acquisition.

3.1 Infiltration of N. benthamiana Leaves

1. Day before infiltration: For each construct, inoculate a colony of *Agrobacterium* in 50 mL of liquid LB medium supplemented with the appropriate antibiotics. Incubate at 28 °C with agitation for 16 h (until reaching an OD_{600} ~ 1) (*see* **Note 2**).
2. Transfer the *Agrobacterium* cultures to centrifuge jars, and collect the cells by centrifugation at 3000 × g for 20 min at room temperature (*see* **Note 3**).
3. During centrifugation, prepare the infiltration buffer. You need 25 mL of infiltration buffer for every 50 mL of *Agrobacterium* culture.
4. Gently resuspend the pellets obtained from centrifugation in 25 mL of infiltration buffer (half the initial volume).
5. Incubate the cultures horizontally in the dark for 90 min with shaking (200 rpm) at 28 °C.
6. Measure the OD_{600} of the cultures (the blank is made with the infiltration solution).
7. Prepare mixtures for infiltration of *Nicotiana* leaves. Each mixture has a final volume of 10 mL, which contains each culture at a final concentration of OD_{600} 0.6 and of the p19 vector at OD_{600} 0.2 (*see* **Note 4**).
8. Infiltrate a minimum of three young *Nicotiana* leaves, starting from the third youngest. Infiltrate the abaxial side of the leaf and in an area away from the middle vein (*see* **Note 5**).
9. Maintain the infiltrated plants in the growth chamber for 24–72 h.

3.2 FRET-FLIM Data Acquisition

1. Cut a 1 cm^2 disc from the infiltrated leaves, avoiding the veins and injection sites. Transfer the samples onto a microscope slide with a drop of water (to avoid sample dehydration), and cover them with a coverslip. Place the slide on a confocal FRET-FLIM microscope system, ensuring that the abaxial side is oriented toward the objective lens.
2. Set the laser at the corresponding wavelength of excitation of both donor and acceptor, and choose HyDx (or R) hybrid detectors (*see* **Note 6**) for both fluorophores. Ensure that there is no crossover between fluorophores (*see* **Note 7**). Specifically for our donor/acceptor example, eGFP is excited at 488 nm and detected between 500 and 540 nm, and mRFP is excited at 543 nm and detected between 580 and 620 nm (Fig. 1). Set the intensity at 10%, speed at 100, and image format at 512 × 512 pixels (*see* **Note 8**). Then, activate FLIM mode (top-left interface) and set the frame repetition to 5 (which means the image will be scanned five times).
3. Start by focusing the sample with a 63× OIL objective. Switch on *Live mode*. Verify the expression levels of the fluorescent protein-tagged proteins of interest. For optimal FRET-FLIM conditions, equal expression levels of all proteins under investigation are ideal (Fig. 2).

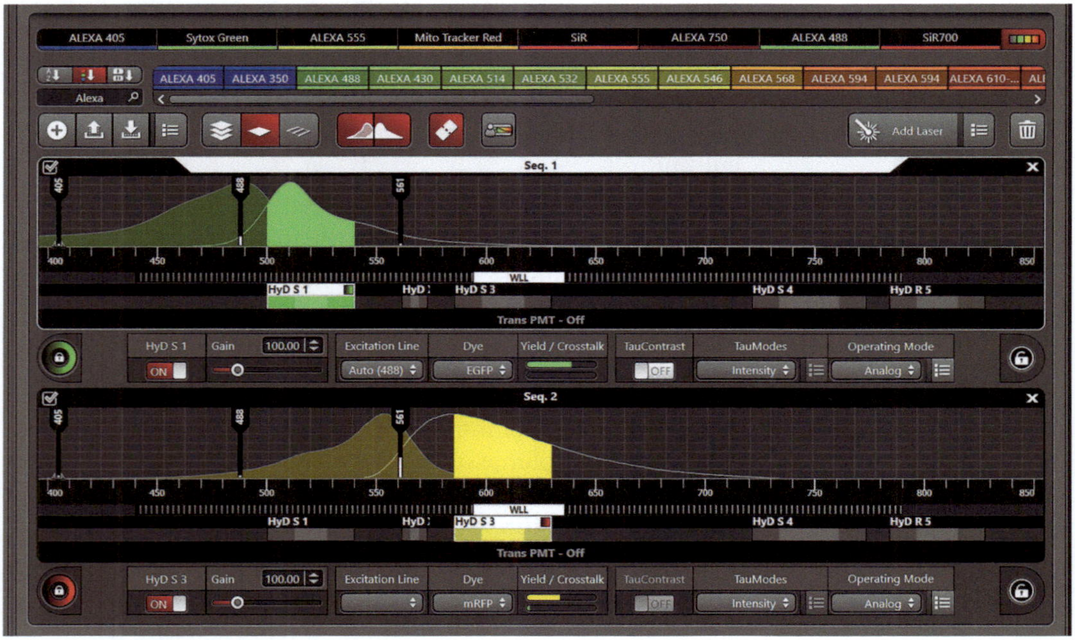

Fig. 1 Screenshot of the settings of fluorophores in Leica LAS X FLIM FCS® display. HyD S detectors have been selected, and eGFP has been configured as the donor (excitation 488 nm; emission 500–540 nm) and mRFP as the acceptor (excitation 543 nm; emission 580–620 nm)

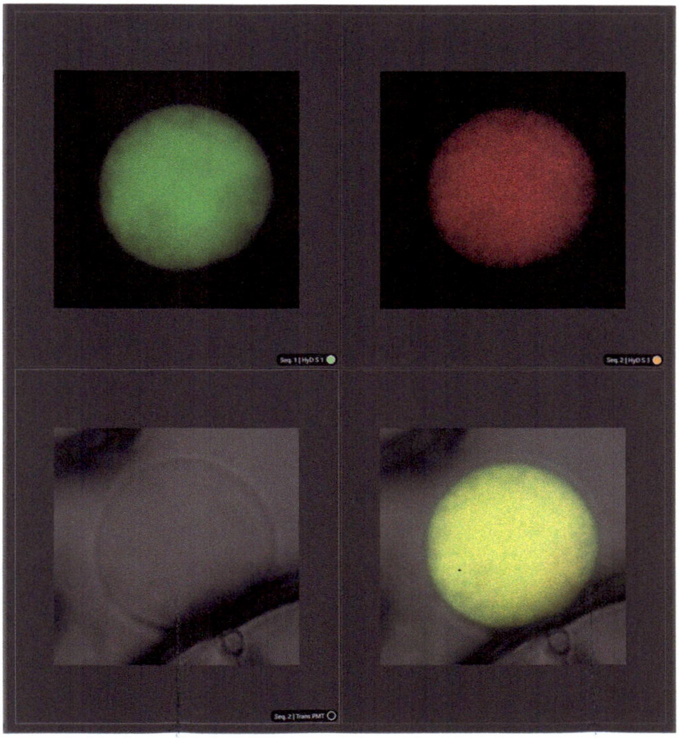

Fig. 2 Screenshot of Leica LAS X FLIM FCS® display showing the localization of equal of the MYBX-eGFP donor (left-top), the bHLHX-mRFP acceptor (right-top), the subcellular compartment (left-bottom), and the merged fields (right-bottom)

4. A specific area of the image, where the interaction is taking place, will be analyzed. For this, a region of interest (ROI) should be drawn using any tool in the top-left panel (*see* **Note 9**).
5. Evaluate the quality of the fitting by different parameters. FLIM data within the specified ROI are represented as fluorescence intensity as a function of time (ns). Therefore, the constant χ^2 (chi-squared) is used to evaluate the fit quality of the exponential decay exhibited by this curve, measuring how closely the data conforms to an exponential function. For instance, χ^2 values close to 1 (0.9 and 1.1) indicate a good fit of the model to the data (Fig. 3) (*see* **Note 10**).
6. Take note of the Lifetime (τ) of the donor alone, once the quality parameters are good.
7. Analyze the FRET-FLIM of protein-protein combination of interest. Take note of Lifetime (τ)—it must be lower than that of the donor alone if a positive interaction occurs (Fig. 4) (*see* **Note 11**).

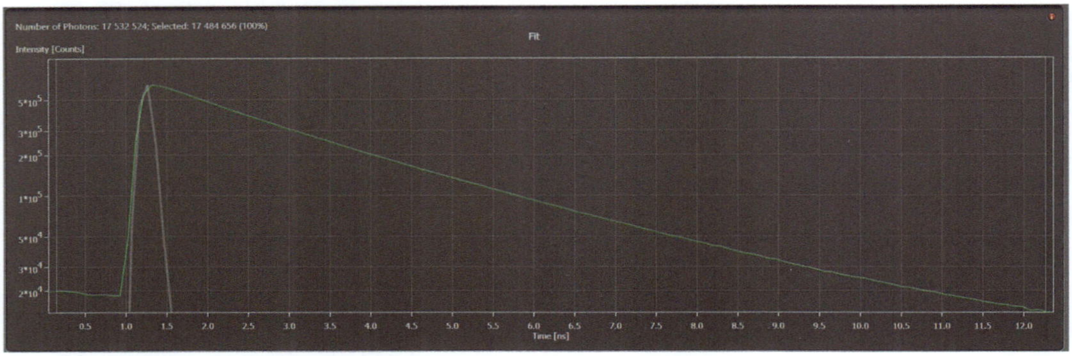

Fig. 3 Screenshot of the Leica LAS X FLIM FCS® interface illustrating the quality of the fitting. The graph displays fluorescence intensity as a function of time (ns), showing a representative decay curve (green) for a single point with an optimal single-exponential fit. The gray curve represents the theoretical fit to the experimental fluorescence decay data

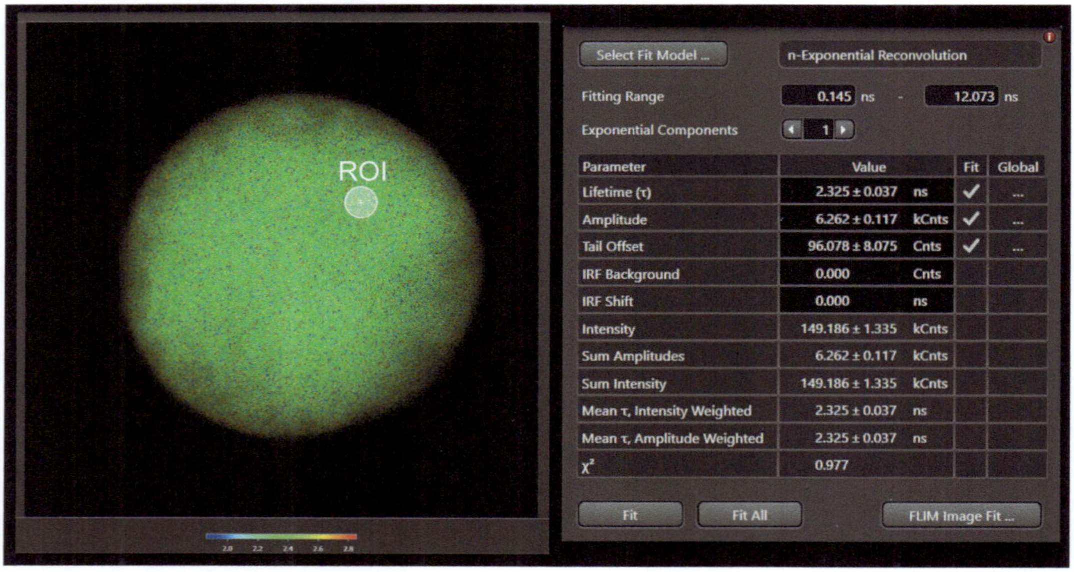

Fig. 4 Screenshot of Leica LAS X FLIM FCS® of an example of FRET-FLIM interaction between two TFs. Left panel, confocal images of a particular ROI showing that MYBX-eGFP interacts with bHLHX-mRFP since the lifetime values for the eGFP/mRFP fusion pair (2.325 $\pm$ 0.037 ns) are lower than those for MYBX-eGFP alone (2.58 $\pm$ 0.04). Right panel, "Fitting parameters" of a certain ROI. χ^2 value is 0.977, ensuring a good fit

8. Analyze at least 10–20 ROIs/sample from at least 3 independent biological samples (plants) per protein-protein combination.

3.3 Analysis and Results Presentation

1. Calculate the average and standard deviation for Lifetime (ns) of positive and negative controls and the interaction of interest. Perform statistical analysis using an unpaired t-test

Table 2
Example of differences in donor fluorescence lifetime (Δ) between interaction and control combinations

Donor	Acceptor	Average eGFP fluorescence lifetime [ns ± SD]	Δ
MYBX-eGFP	mRFP	2.58 ± 0.04	0.28
MYBX-eGFP	bHLHX-mRFP	2.30 ± 0.17	

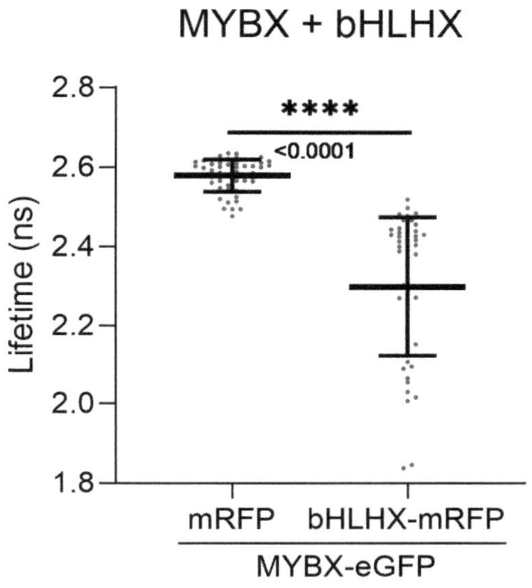

Fig. 5 FLIM lifetime data of eGFP fusion protein as donors in combination with mRFP fusion protein or free mRFP as acceptor. At least 20 measurements were done from independent nuclei for each co-transformation. Asterisks indicate statistical differences as determined by the unpaired t-test (***$p \leq 0.001$)

when comparing two samples or the one-way ANOVA test when comparing three or more.

2. Graph FRET-FLIM data as the Lifetime (τ) of the donor fusion protein (e.g., MYBX-eGFP) vs. the acceptor fusion protein (e.g., bHLHX-mRFP and m-RFP). Example data are shown in Table 2 and Fig. 5.

4 Notes

1. To ensure an accurate assessment of protein interactions, it is essential to clone the proteins of interest with both N-terminal and C-terminal fusion proteins. This approach helps determine whether the interaction is occurring or not. In some cases, only one type of fusion yields a positive result, which may be

attributed to the folded conformations that arise when the protein of interest is fused with a fluorescent protein.

2. It is very crucial to include a culture of the viral silencing suppressor p19.
3. Do not centrifuge at higher speeds to avoid damaging the *Agrobacterium* cells.
4. For a reliable experiment, appropriate expression levels of both donor and acceptor proteins in the plant cells are required. These can be enhanced by changing the OD_{600} of the *Agrobacterium* cultures and the time after infiltration and before visualization. The optimum OD_{600} value is 0.6. The optimal visualization time is 48–72 h after infiltration.
5. It is very important not to generate wounds during the infiltration. If they are generated, do not sample these areas at the time of sampling.
6. Leica's Hybrid detectors (HyDx) are designed for FRET-FLIM.
7. The proper selection of fluorophore pairs for FRET-FLIM is crucial. A common issue in FRET-FLIM occurs when there is a significant spectral overlap of the chosen fluorophore pairs for imaging. Problems can arise when fluorescence from the donor molecule spreads into the acceptor channel and when the acceptor is excited by the laser used to excite the donor. Recommended FRET-FLIM fluorophore pairs for plants include eGFP and mRFP.
8. If the intensity is low, it can be increased, avoiding bleaching. It is recommended to use a laser intensity between 10% and 12%.
9. In the case of using ROIs, select five to ten per sample to ensure that the τ measurements are representative of the entire image. Avoid regions with saturation or no or low expression of donor/acceptor.
10. If the fitting is not adequate ($0.9 < \chi^2 > 1.1$), increase the ROI area until it includes enough intensity information for a good fit.
11. An interaction is considered positive when a 0.2 ns reduction in donor lifetime occurs. A significant lifetime reduction of 0.1 ns may indicate a long-range protein-protein interaction or a non-direct interaction.

Acknowledgments

This research was supported by the Junta de Andalucía (UMA20FEDERJA-093 grant and postdoctoral program, POST-DOC_21_00893 to C.M-P.), a grant (Ref. PID2021-123677OB-

I00 to D. P.), and the Juan de la Cierva postdoctoral fellowship (Ref. JDC2022048407-I to M.F-P.) by the Spanish Ministry of Science, Innovation and Universities.

Conflict of Interest The authors have no conflicts of interest to declare that are relevant to the content of this chapter.

References

1. Blanc-Mathieu R, Dumas R, Turchi L et al (2024) Plant-TFClass: a structural classification for plant transcription factors. Trends Plant Sci 29:40–51. https://doi.org/10.1016/J.TPLANTS.2023.06.023
2. Strotmann VI, Stahl Y (2022) Visualization of in vivo protein-protein interactions in plants. J Exp Bot 73:3866–3880. https://doi.org/10.1093/jxb/erac139
3. Struk S, Jacobs A, Sánchez Martín-Fontecha E et al (2019) Exploring the protein-protein interaction landscape in plants. Plant Cell Environ 42:387–409. https://doi.org/10.1111/pce.13433
4. Bemer M, van Dijk ADJ, Immink RGH, Angenent GC (2017) Cross-family transcription factor interactions: an additional layer of gene regulation. Trends Plant Sci 22:66–80. https://doi.org/10.1016/j.tplants.2016.10.007
5. Brückner A, Polge C, Lentze N et al (2009) Yeast two-hybrid, a powerful tool for systems biology. Int J Mol Sci 10:2763–2788. https://doi.org/10.3390/ijms10062763
6. Phee B-K, Shin DH, Cho J-H et al (2006) Identification of phytochrome-interacting protein candidates in Arabidopsis thaliana by co-immunoprecipitation coupled with MALDI-TOF MS. Proteomics 6:3671–3680. https://doi.org/10.1002/pmic.200500222
7. Spies FP, Perotti MF, Cho Y et al (2023) A complex tissue-specific interplay between the Arabidopsis transcription factors AtMYB68, AtHB23, and AtPHL1 modulates primary and lateral root development and adaptation to salinity. Plant J 115:952–966. https://doi.org/10.1111/tpj.16273
8. Halter T, Imkampe J, Mazzotta S et al (2014) The leucine-rich repeat receptor kinase BIR2 is a negative regulator of BAK1 in plant immunity. Curr Biol 24:134–143. https://doi.org/10.1016/j.cub.2013.11.047
9. Romei MG, Boxer SG (2019) Split green fluorescent proteins: scope, limitations, and outlook. Annu Rev Biophys 48:19–44. https://doi.org/10.1146/annurev-biophys-051013-022846
10. Bracha-Drori K, Shichrur K, Katz A et al (2004) Detection of protein-protein interactions in plants using bimolecular fluorescence complementation. Plant J 40:419–427. https://doi.org/10.1111/j.1365-313X.2004.02206.x
11. McNabb DS, Reed R, Marciniak RA (2005) Dual luciferase assay system for rapid assessment of gene expression in Saccharomyces cerevisiae. Eukaryot Cell 4:1539–1549. https://doi.org/10.1128/EC.4.9.1539-1549.2005
12. Horstman A, Tonaco IAN, Boutilier K, Immink RGH (2014) A cautionary note on the use of split-YFP/BiFC in plant protein-protein interaction studies. Int J Mol Sci 15:9628–9643. https://doi.org/10.3390/ijms15069628
13. Xing S, Wallmeroth N, Berendzen KW, Grefen C (2016) Techniques for the analysis of protein-protein interactions in vivo. Plant Physiol 171:727–758. https://doi.org/10.1104/pp.16.00470
14. Harter K, Meixner AJ, Schleifenbaum F (2012) Spectro-microscopy of living plant cells. Mol Plant 5:14–26. https://doi.org/10.1093/mp/ssr075
15. Piston DW, Kremers G-J (2007) Fluorescent protein FRET: the good, the bad and the ugly. Trends Biochem Sci 32:407–414. https://doi.org/10.1016/j.tibs.2007.08.003
16. Liu J, Huang Y, Kong L et al (2020) The malectin-like receptor-like kinase LETUM1 modulates NLR protein SUMM2 activation via MEKK2 scaffolding. Nat Plants 6:1106–1115. https://doi.org/10.1038/s41477-020-0748-6
17. Long Y, Stahl Y, Weidtkamp-Peters S et al (2018) Optimizing FRET-FLIM labeling conditions to detect nuclear protein interactions at native expression levels in living arabidopsis roots. Front Plant Sci 9:639. https://doi.org/10.3389/fpls.2018.00639

18. Kriechbaumer V, Botchway SW, Hawes C (2016) Localization and interactions between Arabidopsis auxin biosynthetic enzymes in the TAA/YUC-dependent pathway. J Exp Bot 67: 4195–4207. https://doi.org/10.1093/jxb/erw195

19. Kriechbaumer V, Botchway SW, Slade SE et al (2015) Reticulomics: protein-protein interaction studies with two plasmodesmata-localized reticulon family proteins identify binding partners enriched at plasmodesmata, endoplasmic reticulum, and the plasma membrane. Plant Physiol 169:1933–1945. https://doi.org/10.1104/pp.15.01153

20. Stahl Y, Grabowski S, Bleckmann A et al (2013) Moderation of Arabidopsis root stemness by CLAVATA1 and ARABIDOPSIS CRINKLY4 receptor kinase complexes. Curr Biol 23:362–371. https://doi.org/10.1016/j.cub.2013.01.045

Part III

Analysis and Inference of Transcriptional Gene Regulatory Networks

Chapter 12

MINI-EX Version 2: Cell-Type-Specific Gene Regulatory Network Inference Using an Integrative Single-Cell Transcriptomics Approach

Jasper Staut, Nicolás Manosalva Pérez, Thomas Depuydt, Klaas Vandepoele, and Svitlana Lukicheva

Abstract

Understanding and predicting cell-type-specific gene regulatory networks (GRNs) is essential for unraveling the complex interactions between transcription factors (TFs) that modulate the expression of target genes and control diverse biological processes in multicellular organisms. MINI-EX (Motif-Informed Network Inference based on single-cell EXpression data) is an integrative tool tailored for identifying cell-type-specific GRNs in plants. Leveraging single-cell transcriptomics data, MINI-EX constructs expression-based networks and integrates TF motif information to produce GRNs with increased accuracy. Furthermore, it assigns regulatory modules to distinct cell types and prioritizes candidate regulators by employing a strategy that encompasses network centrality measures, functional annotations, and expression specificity. Taken together, MINI-EX offers a powerful approach to identify cell-type-specific transcriptional cascades and enhance our understanding of TF functions in plant biology. Here, we discuss recent advancements in the tool's latest version and explain how single-cell GRNs can be identified for non-model species lacking TF motif information. Additionally, we provide a comprehensive guide to use MINI-EX, covering the entire pipeline from preparing input files starting from a single-cell experiment, over configuring parameters, to interpreting output data.

Key words Single-cell RNA-seq, Gene regulatory network, Transcription factors, Systems biology

1 Introduction

Recently, single-cell technologies have been increasingly used to gain insight into complex biological mechanisms [1–4]. Zooming in on the level of the single cell allows tracking dynamic cellular processes, study rare cell populations, and differentiate between tissue-specific responses to stress or hormone treatment experiments [1, 5]. While initial single-cell studies in plants focused on model plant *Arabidopsis thaliana* [4, 6–10], the plant single-cell field has since been extended to important crop species such as

José M. Franco-Zorrilla and Carmen Martín-Pizarro (eds.), *Plant Transcription Factors: Methods and Protocols*, Methods in Molecular Biology, vol. 2985, https://doi.org/10.1007/978-1-0716-4972-5_12,

maize, rice, and tomato [11–14]. With the new opportunities the single-cell technology brings also comes the computational challenge to represent, integrate, and leverage the information these datasets encompass, with the ultimate goal to extract novel, biological knowledge from them. With this objective in mind, gene regulatory network (GRN) inference can offer a suitable approach to integrate data and represent the regulatory relationships between genes, facilitating the identification of key regulators [15–17]. The GRNs consist of nodes, representing genes, and directed edges between them, denoting a regulatory relationship between a transcription factor (TF) to a target gene (TG) [18]. TFs are proteins that bind to specific DNA sequences, called motifs, to regulate the expression of a TG. Together, one TF and its TGs are referred to as a regulon.

In a recent publication, MINI-EX, a tool for cell-type-specific GRN inference and prioritization of candidate regulators in plants, was presented [19]. It is designed to integrate single-cell RNA sequencing data (scRNA-seq) with TF binding motif information, inferring high-quality GRNs for each cell type, and prioritize most relevant regulators using network statistics and gene ontology (GO; a controlled vocabulary describing current knowledge about the function and location of genes products) information. Both in its initial release and subsequent comparisons, MINI-EX has been shown to outperform the previously developed GRN inference tool SCENIC and is gaining interest within the plant single-cell community [19–21]. In the recent release of MINI-EX version 2.0, its algorithm, visualizations, and documentation have been further improved. In addition, new features were added such as the possibility to omit the integration of TF binding motif information, which provides a quick and straightforward way to run MINI-EX for non-supported species without the need for additional preprocessing of motif data for those species. In this chapter, we offer a practical guide on how to use MINI-EX for cell-type-specific GRN inference. We will cover the requirements for running the algorithm, briefly explain its basic principles, show how to prepare the configuration and input files, and demonstrate how to run the pipeline. Finally, we discuss all output files and provide an example of how the results from MINI-EX can be interpreted to advance understanding of biological processes.

2 Materials

2.1 Hardware and Software

MINI-EX, implemented as a Nextflow [22] pipeline and distributed with a Singularity [23] container, can be run on any POSIX compatible system (such as Linux, OS X, etc.) and requires only Nextflow and Singularity installations for its execution. Since MINI-EX is designed to handle large datasets, it is recommended for use on high performance computing (HPC) clusters. Version 2.2 of MINI-EX (used in this chapter) requires approximately 17 hours to run on the Wendrich et al. (2020) [4] dataset discussed in this chapter (and only 30 minutes when using a precomputed GRNBoost2 output file) and uses a maximum of 20 Gb of RAM (detailed memory requirements per process can be found in the config file available on GitHub). The example dataset provided on the GitHub repository can be run on a personal computer and takes around 18 minutes for the full run and around 3 minutes with a precomputed GRNBoost2 output file.

2.2 Expertise

MINI-EX execution begins with information extracted from a processed single-cell RNA sequencing (scRNA-seq) dataset. Therefore, we assume that the user possesses the capability to conduct a conventional single-cell processing analysis. Further in the chapter, we provide detailed instructions on how to extract MINI-EX input files from a Seurat object; an example of processing scRNA-seq data using the Seurat library can be found in [24]. However, users are free to use alternative tools, provided they can extract the required MINI-EX input files in the correct format.

2.3 Input Files

To execute MINI-EX, the four following files, derived from processing an experimental scRNA-seq dataset, are required:

1. An expression matrix, which provides the expression levels of each gene (row) in each cell (column)
2. A list of marker genes per cluster, in Seurat's findAllMarkers output format
3. A file containing the cluster identity of each cell
4. A file containing the cluster annotations

MINI-EX currently provides support for four plant species: *Arabidopsis thaliana*, *Oryza sativa* (rice), *Zea mays* (mays), and *Solanum lycopersicum* (tomato). For these species, the user is able to fully utilize all MINI-EX features. For users interested in running MINI-EX on plant species not currently supported, it is possible to disable the motif analysis step (*see* **Note 1**). In this case, users only have to provide a list of TFs of the species of interest. Additional files may be generated (see Subheading 3.5 on how to do this) to leverage the full capabilities of MINI-EX, but are not required.

2.4 Example Dataset

To illustrate the execution of MINI-EX and the evaluation of its output files, we will use the Wendrich Arabidopsis scRNA-seq root dataset published in [4]. The corresponding Seurat file is available for download from the MINI-EX GitHub page at the following link: https://github.com/VIB-PSB/MINI-EX/tree/main.

In summary, the input FASTQ files were initially processed using CellRanger v6.1.1. The resulting files were then employed to create the Seurat object using the Seurat R package v4.0.1. We set upper (2%) and lower (25%) thresholds on library size, merged the samples, and performed normalization using Seurat's SCTransform function [25], with the number of variable features set to 7000. For UMAP creation, we used 70 principal components with a resolution of 1.2. After processing, this dataset contains 11,456 cells, 21,675 expressed genes, and 38 clusters, encompassing 15 tissues.

3 Methods

3.1 Basic Principles

MINI-EX utilizes scRNA-seq data as input to infer cell-type-specific GRNs and prioritize relevant regulators. The MINI-EX algorithm consists of four general steps (Fig. 1). In **step 1**, a coexpression-based GRN is generated using GRNBoost2 [26] yielding an initial expression-based network. In **step 2**, this network is filtered by retaining only putative TF-TG interactions if the TF binding motif is found to be enriched (through motif mapping and statistical analysis) in the regulatory region (i.e., 5000 bp upstream and 1000 bp downstream of the gene's translation start/stop site and its introns) of a TG, and the regulon is enriched for the TF motif. This motif-filtered network is then used in **step 3** where the expression of TFs and TGs in each cluster is used as a last filtering step, resulting in multiple cluster-specific GRNs. Finally, **step 4** integrates network statistics and optional gene ontology (GO) information (functional annotations of genes) to rank the regulons in each cell type on their importance.

3.2 MINI-EX 2.0 and Later

In this subheading, we will briefly outline the changes that were made between the original MINI-EX release (version 1.0) and its later iterations. Given that this book chapter describes the use of MINI-EX 2.2, we want to clearly specify the differences between the original MINI-EX 1.0 and the version that is used here. The most relevant changes are listed below.

Version 2.0

1. MINI-EX can be easily run for any non-supported species by adding the possibility to omit the motif enrichment step (see **step 2**, Fig. 1). Doing so lowers the precision of the resulting networks (*see* **Note 1** for more details) but removes the need for preprocessed motif information that is only provided for supported species.
2. Support for tomato genome version SL4.0/ITAG4.0.

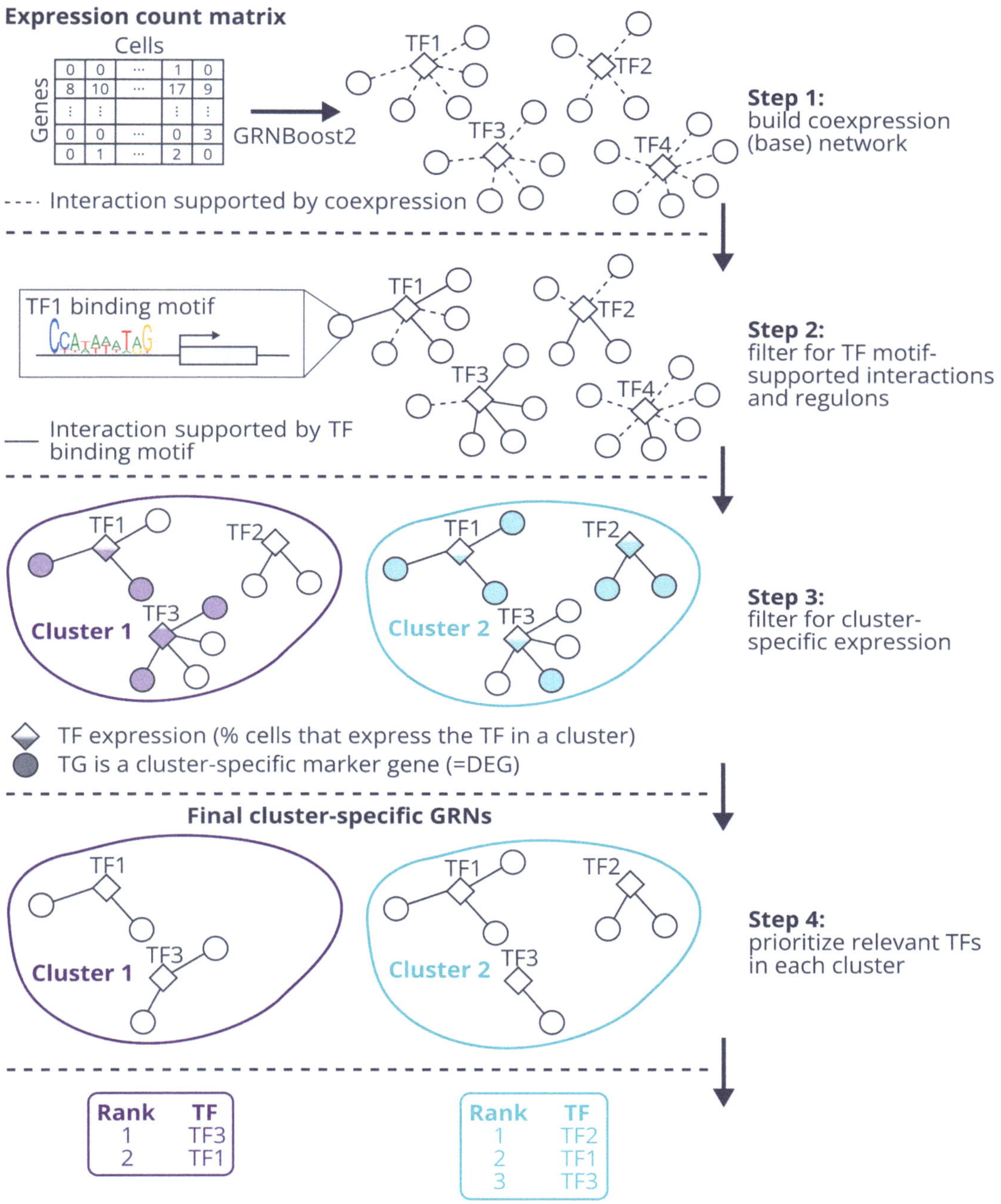

Fig. 1 Generalized MINI-EX workflow. The main steps of the MINI-EX workflow are illustrated, showing a collection of regulons, each consisting of a transcription factor (TF; shown as diamonds) and its putative target genes (TGs; shown as circles), together forming a gene regulatory network (GRN). **Steps 1** and **2** result in a single GRN for the full dataset that forms the precursor for **step 3**, which operates in a cluster-specific manner, yielding a unique GRN for each cluster of the single-cell input data. In **step 4**, TFs are prioritized based on network statistics and optional gene ontology terms (not shown)

3. Addition of a new regulator heatmap visualization, called "regmap." This visualization allows inspecting regulator importance over different cell-type clusters and/or developmental trajectories.
4. The procedure to calculate enrichment statistics is modified to conform with community guidelines [27, 28]. With the aim of improving performance, the enrichment procedure is modified which results in differences in output between version 1.0 and 2.0 or later (*see* **Note 2** for more details).
5. Other minor improvements are implemented such as bug fixes, improved documentation, and input checks.

Version 2.1

6. Support for maize genome version AGPv5.
7. Addition of computationally inferred GO terms to the gene-GO association files [29] provided for supported species, except for *A. thaliana* (*see* **Note 3** for more details).

Version 2.2

8. Added high-throughput evidence codes to the Arabidopsis gene-GO association file.
9. Improvements in the enrichment algorithm.
10. Added log file summarizing statistics on the input datasets and intermediate results in the workflow.
11. Improved layout of the generated figures.
12. Gene alias file is made optional.

3.3 The MINI-EX Configuration File

Before running MINI-EX, the user first needs to prepare the configuration (config) file, which contains all settings that the user needs to specify before starting the pipeline. This includes the path of all input files and the output folder, all parameters, as well as memory settings. In this subheading, we describe the general setup of the config file. In Subheadings 3.4 and 3.5, we describe how to generate the input files and Subheading 3.6 covers the MINI-EX parameters. Changing the memory settings is covered in Subheading 3.7, which also explains how to run the pipeline.

The config file starts with two smaller sections. The first section specifies the job scheduler or executor and is covered in Subheading 3.7 about how to run the pipeline. The second section relates to the Singularity container that MINI-EX uses to handle dependencies and should not be adapted by the user. The following section specifies input files and parameters. Figure 2 shows the MINI-EX algorithm workflow, including all input files and parameters that are to be set in the config file. Input files that are required by MINI-EX fall in two categories: input files to be provided by the user and files shipped with MINI-EX for supported species (see Fig. 2). The user-

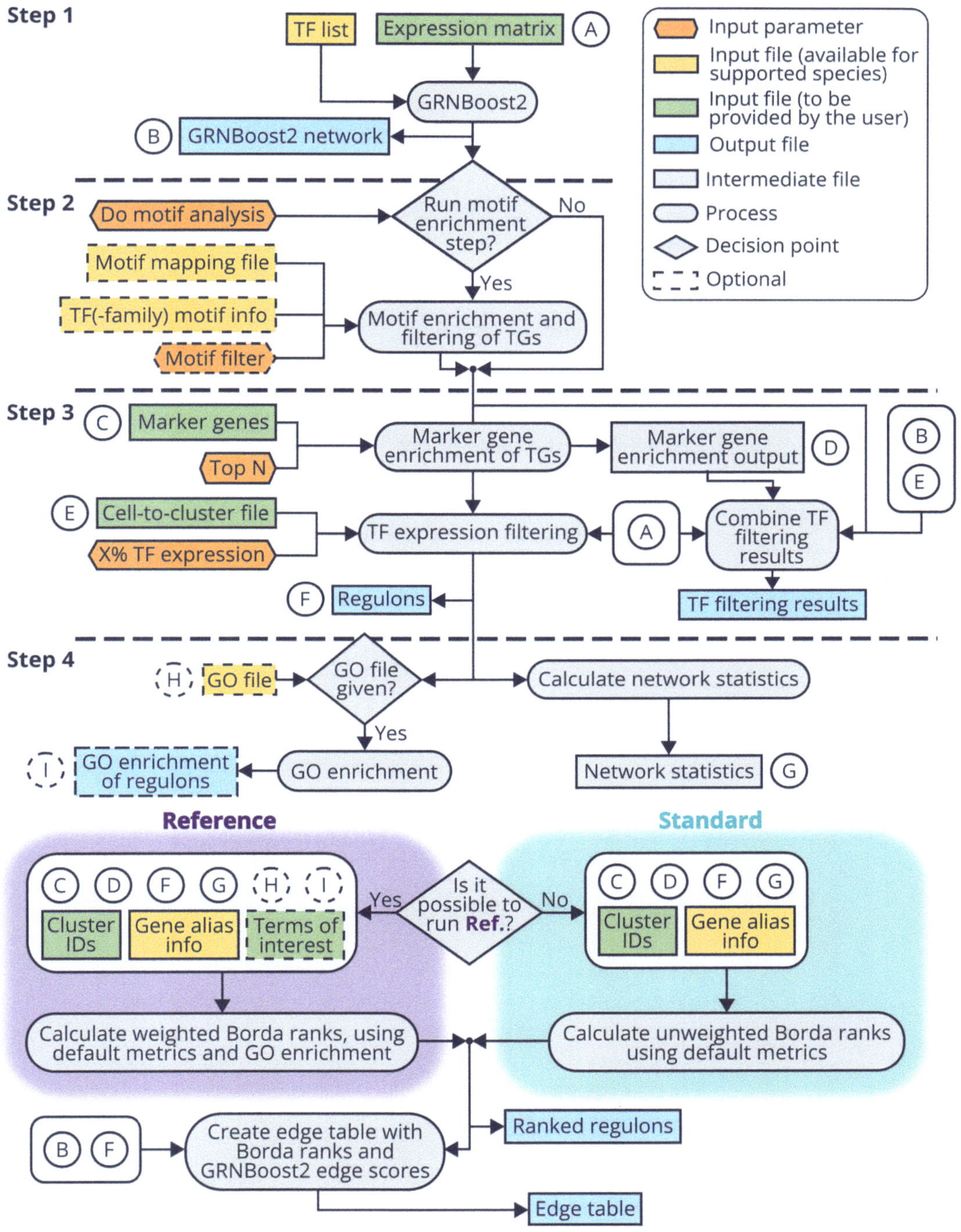

Fig. 2 MINI-EX algorithm overview. Input files and parameters are used in a sequence of processes, producing several output files. Different options (shown as diamonds) are available for the user, corresponding with decision points in the workflow. Processes and files related to visualizations are not included in this schema

provided files are specified first in the config file. These files are mandatory input and include four files derived from a single-cell dataset:

```
expressionMatrix = "$baseDir/example/INPUTS/*_matrix.tsv"
markersOut = "$baseDir/example/INPUTS/*_allMarkers.tsv"
cellsToClusters = "$baseDir/example/INPUTS/*_cells2clusters.
tsv"
clustersToIdentities = "$baseDir/example/INPUTS/*_identities.
tsv"
```

The wildcard character "*" in the file names above allows Nextflow to recognize input files with a different prefix (followed by an underscore) as coming from different datasets and run them in parallel (see Subheading 3.7). Files of the same dataset must always share the same prefix.

The second type of files includes generic data related to the species of interest. These files are available on the MINI-EX GitHub repository for the four supported species but must be provided by the user for the non-supported species. In this case, only the list of transcription factors is mandatory; the remaining files should be specified if the user wants to fully utilize all of MINI-EX's features:

```
tfList = "$baseDir/data/ath/ath_TF_list.tsv"
geneAliases = "$baseDir/data/ath/ath_gene_aliases.tsv"
infoTf = "$baseDir/data/ath/ath_TF2fam2mot.tsv"
featureFileMotifs = "$baseDir/data/ath/ath_2021.1_motifMap-
ping.out.gz"
goFile = "$baseDir/data/ath/ath_full_BP_expcur_ext_names.tsv"
```

After the input files, parameters are defined:

```
doMotifAnalysis = true
termsOfInterest = "$baseDir/example/INPUTS/GOsIwant.txt"
motifFilter = "TF-F_motifs"
topMarkers = "700"
expressionFilter = "10"
topRegulons = "150"
```

The last section of the config file sets the memory for the different processes and will be discussed in Subheading 3.7.

3.4 Preparing Input Files from a scRNA-seq Experiment

This subheading outlines the preparation of the four essential files required for running MINI-EX from a scRNA-seq experiment. In the examples below, we show how to extract the required input from a scRNA-seq experiment processed with the Seurat R package. It is important to note that all input files, except for the marker genes, should have unique row names and data should be stored in tab-delimited format without quotation marks. In the examples below, we will use the Seurat object available at https://zenodo.org/records/10423816?token=eyJhbGciOiJIUzUxMiJ9.eyJpZCI6ImUzZmQyYmRhLWUwMTItNGY1My04NDI4LTk0MThlZjRhYzE1OSIsImRhdGEiOnt9LCJyYW5kb20iOiJhZTM2ZjI2ZjA5YTQzMWQ2NjIxZDQwNTEwMzQxNTExZSJ9.Ge8ym9h2nTvFZPreJz85PBrelLaiGXh3b_CjOW9xzLod-Bi89ZVlWczrW3MlMVWGX55rOKEcP70PI8n2qegQEw.

1. **Loading the Seurat object**. In the examples below, we assume that the Seurat object has been loaded and stored in the variable "seurat.obj," as demonstrated below. MINI-EX requires that all scRNA-seq-related input files from the same dataset begin with the same prefix; we therefore store the prefix of our dataset in the variable "out.prefix."

```
library(Seurat)
seurat.obj <- readRDS(file = "/input_dir/Wendrich.rds")
out.prefix <- "athWendrich"
out.dir <- "/output_dir"
```

2. **Expression matrix**: This file provides the expression levels of each gene in each cell. The format for this input file should be a tab-separated file with gene identifiers as rows and cell identifiers as columns. Specifically, the cell in row X and column Y should contain the expression level of the gene X in cell Y. While we recommend providing the original (non-normalized) expression levels (counts), alternative approaches may be considered (*see* **Note 4**). The expression matrix can be generated using the following code.

```
write.table(as.matrix(seurat.obj@assays$RNA@counts),
file = sprintf("%s/%s_matrix.tsv", out.dir, out.prefix),
sep = "\t", quote = FALSE)
```

3. **Marker genes**: This is a list of differentially expressed genes (markers) between a cluster and all other clusters in the scRNA-seq dataset, for each cluster in the dataset. Only upregulated markers are allowed (*see* **Note 5**) for MINI-EX. This file should follow the format of Seurat's "FindAllMarkers" function and can be generated using the code below.

```
markers <- FindAllMarkers(seurat.obj, only.pos = TRUE)
write.table(markers, sprintf("%s/%s_allMarkers.tsv", out.dir,
out.prefix), sep = '\t' , quote = FALSE)
```

4. **Cluster identities of cells**: This is a tab-separated file with cells in the first column and their corresponding assigned clusters in the second column.

```
cells2cluster <- FetchData(seurat.obj, vars = 'ident')
write.table(cells2cluster, sprintf("%s/%s_cells2clusters.
tsv", out.dir, out.prefix), sep='\t', quote = FALSE, col.names
= FALSE)
```

5. **Cluster annotation**: This is a tab-separated file with cluster identities (matching the previous file) in the first column and their annotations in the second column. The user can also add an optional third column to configure the cluster order in the regmap figure (see Subheading 3.8), where the column index of each cluster in the plot is added as a third column. This file can be created manually. Cluster annotations should not contain underscore characters. The first lines of the cluster annotation file for our example dataset are provided below; the full file can be downloaded from https://zenodo.org/records/10423816?token=eyJhbGciOiJIUzUxMiJ9.eyJpZCI6ImUzZmQyYmRhLWUwMTItNGY1My04NDI4LTk0MThlZjRhYzE1OSIsImRhdGEiOnt9LCJyYW5kb20iOiJhZTM2ZjI2ZjA5YTQzMWQ2NjIxZDQwNTEwMzQxNTExZSJ9.Ge8ym9h2nTvFZPreJz85PBrelLaiGXh3b_CjOW9xzLod-Bi89ZVlWczrW3MlMVWGX55rOKEcP70PI8n2qegQEw.

16	pericycle
24	Xylem
9	Epidermis
28	Trichoblast
0	LRC

3.5 Preparing Input Files for Non-supported Species

Users planning to run MINI-EX on the supported species only need the four files related to a scRNA-seq experiment. For non-supported species, in addition to these files, a list of TFs specific to the species must also be provided. Additionally, users may choose to include further files based on the specific MINI-EX features required for their analysis. The description of these additional files is provided below. It is important to verify that the gene identifiers used in these files match those used in the scRNA-seq experiment.

1. **List of TFs** (mandatory): This file is essential for inferring the initial GRN as depicted in Fig. 1, **step 1**. This file only has one column corresponding to gene identifiers of transcription

factors. The list of transcription factors of the species of interest can be downloaded from PlantTFDB (http://planttfdb.gao-lab.org/index.php) [30].

2. **Motif mapping** (optional): This file is necessary to execute **step 2** of MINI-EX, which involves filtering the original GRN using TF motif-supported information. Detailed instructions on generating this file can be found in [31, 32] (*see* **Note 6**).
3. **Transcription factor family and motifs** (optional): This is the second requirement for running **step 2** of MINI-EX. It establishes the connections between each TF, its family, and the availability of motif information. This file is tab-delimited and includes four columns:
   (a) TF identifier
   (b) TF family
   (c) "Y" if motif information is available, "N" otherwise
   (d) Motif information

   The data for the first two columns can be downloaded from PlantTFDB (http://planttfdb.gao-lab.org/index.php) [30], while TF motif information can be obtained from JASPAR (https://jaspar.genereg.net/downloads/) [33] and CisBP (http://cisbp.ccbr.utoronto.ca/bulk.php) [34]. An example of the TF family and motifs file is provided below, with the first lines of the file available on the MINI-EX GitHub for rice. Note that for the TF on the last line, no motif information is available.

```
LOC_Os01g04750  RAV  Y  M06692,M01648,M06693,M09748,
                        M09749
LOC_Os01g04800  RAV  Y  M06692,M01648,M06693,M09748,
                        M09749
LOC_Os05g47650  RAV  Y  M06692,M01648,M06693,M09748,
                        M09749
LOC_Os01g49830  RAV  Y  M06692,M01648,M06693,M09748,
                        M09749
LOC_Os01g01430  NAC  N
```

4. **GO file** (optional): This file is only required if the user has specified a list of terms of interest. It serves to prioritize regulons based on these user-defined terms during the **step 4**, with the last column (GO term description) used for filtering relevant GO terms. In case no terms of interest are provided, the GO file can still be provided to generate GO enrichment information about the regulons' TGs, but this will not affect the final GRNs or regulon prioritization. The GO file lists gene-

GO associations for biological processes in a tabular format, consisting of four columns:

(a) GO term identifier
(b) Gene identifier
(c) Evidence code
(d) GO term description

The contents of this file must be obtained from the data provider for the species of interest. To ensure accurate results in enrichment analysis, we recommend performing parental term propagation, which may be accomplished using various available tools (e.g., GOATOOLS) [35]. For non-model species, which generally lack extensive annotations, we recommend retaining all the information. However, for more extensively studied species, it can be advantageous to filter annotations based on experimental evidence codes (the list of codes to keep can be found on the Gene Ontology website: https://geneontology.org/docs/guide-go-evidence-codes/) (*see* **Note 3**). An example is shown below using the gene-GO association file provided in the MINI-EX GitHub for rice.

GO:0000902	LOC_Os01g01030	ISO	cell morphogenesis
GO:0008152	LOC_Os01g01030	IEA	metabolic process
GO:0009653	LOC_Os01g01030	ISO	anatomical structure morphogenesis
GO:0009826	LOC_Os01g01030	ISO	unidimensional cell growth
GO:0009932	LOC_Os01g01030	ISO	cell tip growth

5. **List of gene aliases** (Optional): This file allows users to display gene aliases instead of gene identifiers on the output files provided by MINI-EX. It is a tab-formatted file with two columns: gene identifier and gene alias. An example is shown below using the first lines of the gene alias file provided in the MINI-EX GitHub for rice. If no gene aliases are available, MINI-EX will display the original gene identifiers in the output files.

LOC_Os07g46460	ABC1
LOC_Os07g46460	OsFd-GOGAT
LOC_Os07g46460	SPL32
LOC_Os07g46460	ES7
LOC_Os07g39750	AchE

3.6 Setting MINI-EX Parameters

The behavior of MINI-EX depends on several parameters to be set by the user (*see* **Note 7** on using the default parameters). Here, all parameters will be described as well as the rationale behind tuning those parameters provided in accompanying notes. Figure 2 illustrates how these settings are integrated in the MINI-EX algorithm.

1. **Do motif analysis**: The default value is "true," resulting in the conventional way of running MINI-EX including a motif enrichment step (see **step 2**, Fig. 2). Setting this parameter to "false" will omit the motif enrichment step in the algorithm. Disabling this option is not recommended because it will reduce the precision of the resulting networks (*see* **Note 1** for more details). However, by omitting this step, no motif or TF family information is required anymore. This allows running MINI-EX for any non-supported species, without the need to generate the additional files described in Subheading 3.5.
2. **Terms of interest** (Optional): This is an optional file that the user can provide, affecting how the regulons are ranked. This file contains a set of terms that are used to retrieve GO terms (*see* **Note 8** on how this is done) related to the research question of the user. When this file is not provided (set to "null" in the config file), MINI-EX will use its standard procedure to rank regulons, based only on network statistics and cluster specificity (*see* **Note 9** for more details). When the user does provide this file, MINI-EX will use GO information to give a better rank to regulons that have an overrepresentation of the GO terms of interest among their target genes (*see* **Note 10** for more details).
3. **Motif filter**: This parameter can either have the value "TF_motifs" or "TF-F_motifs" (default) and affects how strict the motif enrichment filtering is in step 2 (see Fig. 2). Setting it to "TF_motifs" results in filtering for regulons supported by the specific binding motif(s) of a regulon's TF, whereas "TF-F_motifs" (default) results in a more liberal filtering, keeping regulons supported by any TF binding motif within the TF protein family. As a result, using "TF-F_motifs" allows keeping regulons controlled by TFs that lack motif information (*see* **Note 11** for more details) and thus increases the number of TFs that can be studied.
4. **Top markers**: This is the number of marker genes for each cluster that are selected for the marker gene enrichment and filtering in **step 3a** of MINI-EX. This parameter can strongly affect the size of the final regulons with higher numbers leading to bigger regulons (*see* **Note 12** for more details).
5. **Expression filter**: This is the percentage of cells in a single-cell cluster that shows expression of a TF, which is used as the cutoff in **step 3b** (Fig. 2) to filter out regulons with a TF that

is not expressed in that cluster. The higher this number, the more stringent the filtering for what is considered an expressed TF (*see* **Note 13** for more details).

6. **Top regulons**: This parameter corresponds with the top number of regulons per cluster that are displayed in the visualizations of the MINI-EX output. It solely affects what is displayed in the output figures. The GRNs and ranking remain the same.

3.7 Running the Pipeline

Running MINI-EX involves setting up the config file ("miniex.config") with all the necessary input paths and parameters and executing the following command:

```
nextflow -C miniex.config run miniex.nf
```

MINI-EX enables the concurrent processing of multiple scRNA-seq datasets. This can be achieved by incorporating the wildcard character "*" in the names of the four mandatory input files, as explained in Subheading 3.3. For example, using "*_allMarkers.tsv" will find all files ending with "_allMarkers.tsv" but starting with different dataset names (i.e., different prefixes). Please note that this feature only allows running MINI-EX on multiple datasets of the same species, as the remaining files and parameters, including the terms of interest, remain fixed in the config file.

MINI-EX is primarily designed for HPC clusters and is set to run with the job scheduler SLURM by default:

```
executor {
name = 'slurm'
queueSize = 5
}
```

Before initiating MINI-EX, users should verify which job scheduler (executor) is available on their system and adjust the "name" parameter within the "executor" scope accordingly (*see* **Note 14**). If the user does not intend to run MINI-EX on an HPC cluster, they can set the "name" parameter to "local." The list of acceptable values for this parameter can be found in the Nextflow documentation: https://www.nextflow.io/docs/latest/executor.html.

As shown above, the default value for the "queueSize" parameter is 5, meaning that Nextflow is only allowed to run five tasks in parallel. This value can be adjusted by the user as needed and must be set to 0 if the user does not want to set a limit.

Depending on the user's dataset, some steps may require more or less resources. Default memory requirements are specified in the config file for each process; however, for some datasets they may not be sufficient. If MINI-EX fails at a particular step either with a

"MemoryError" or without any specific error message provided by MINI-EX, it is highly probable that the process was killed due to insufficient memory. This can be corrected by increasing the "memory" parameter of the corresponding dataset in the config file:

```
withName: run_grnboost {
memory = '20 GB'
cpus = 5
}
```

As constructing the GRN (**step 1**), performed by GRNBoost2 [26], is one of the most resource-intensive steps, it is possible to store the GRNBoost2 output for reuse in subsequent runs on the same dataset. This can be accomplished using the "grnboostOut" parameter. For the first run, this parameter must be set to "null," indicating that the base GRN needs to be computed. For subsequent runs, this parameter can be set to the path of the GRNBoost output file:

```
grnboostOut = "/$baseDir/example/OUTPUTS/GRNBoost2_output/
*_grnboost2.tsv"
```

3.8 The MINI-EX Output Files

When MINI-EX has been run successfully, output files are stored in separate folders, named GRNBoost2_output, GOenrichment_output, regulons_output, and figures. In this subheading, the different output files contained in these folders will be described along with their interpretation. We first discuss the ranked regulons file and the regulons file, as these can be considered the two most important output files of MINI-EX, followed by a discussion of the output figures.

1. **Ranked regulons file** (*_rankedRegulons.xlsx or *_rankedRegulons.tsv): This file, provided in different file formats for user convenience, contains metadata and statistics on the final regulons inferred by MINI-EX, including their relevance ranking.
   (a) **TF information** (columns 1–5; Table 1): Column 1 contains the gene ID of the TF, column 2 is its alias (i.e., gene symbol), and column 3 shows whether the TF is already known (GO information is available for it) and whether it is a relevant TF (at least one of the associated GO terms matches a term of interest given by the user; *see* Subheading 3.5 and **Note 8**). Columns 4 and 5 show the GO terms associated with the TF and their description, respectively.
   (b) **Cluster-related information** (columns 6–9; Table 2): Column 6 contains the cluster label and column 7 the annotation of that cluster (as given in the cluster

Table 1
Ranked regulons result table, columns related to TF information

TF	Alias	hasTFrelevantGOterm	GOterm	GOdescription
AT3G58710	WRKY69/ ATWRKY69/ ABT4	unknown_TF	–	–
AT5G26660	MYB86/ ATMYB4-old/ ATMYB86/ ATMYB4	known_TF	GO:0006139, GO:0006351, ...	Nucleobase-containing compound metabolic process, DNA-templated transcription
AT5G16600	MYB43/ AtMYB43	known_TF	GO:0006082, GO:0006139, ...	Organic acid metabolic process, nucleobase-containing compound metabolic process
AT3G61850	DAG1	known_TF	GO:0006139, GO:0006351, ...	Nucleobase-containing compound metabolic process, DNA-templated transcription
AT5G18270	ANAC087	unknown_TF	–	–

Table 2
Ranked regulons result table, columns with cluster-related information

cluster	celltype	isTF_DE	totRegInCluster
trichoblast_Cluster_28	trichoblast	1	132
atrichoblast_Cluster_37	atrichoblast	1	130
procambium_Cluster_17	procambium	1	151
procambium_Cluster_17	procambium	1	151
LRC_Cluster_33	LRC	1	129

annotation input file). Column 8 shows whether the TF of the regulon is a marker gene for that cluster, and column 9 gives the total number of regulons that were retained for that cluster.

(c) **Regulon importance metrics** (columns 10–14; Table 3): Column 10 contains the number of TGs in the regulon. Column 11 contains the cluster specificity, which corresponds to the q-value of enrichment between the

Table 3
Ranked regulons result table, columns with regulon statistics for ranking

#TGs	qval_cluster	out-degree	closeness	betweenness
673	1,4E-254	0,821734	0,126227	0,006404
668	0	0,817625	0,131409	0,005853
683	8,8E-267	0,825877	0,147863	0,00381
683	7,6E-157	0,825877	0,147863	0,003663
694	0	0,864259	0,136936	0,002872

Table 4
Ranked regulons result table, columns with regulon gene ontology enrichment information

GO_enrich_qval	GO_enrich_term	GO_enrich_desc	#TGs_withGO
4,35E-30	GO:0010053	root epidermal cell differentiation	38
2,75E-12	GO:0022622	root system development	42
3,62E-06	GO:0022622	root system development	31
3,62E-06	GO:0022622	root system development	31
0,014191	GO:0022622	root system development	21

regulons' TGs and the marker genes of that cluster (calculated in **step 3a** of MINI-EX; Fig. 2). Columns 12–14 contain three network centrality metrics—out-degree, closeness, and betweenness—which measure how central the TF is embedded within the network structure (*see* **Note 15**).

(d) **Enrichment of TGs for GO terms of interest** (columns 15–18; Table 4): These four columns are only present when the input file with terms of interest was provided by the user. The column "GO_enrich_qval" contains the q-value of the most significantly enriched GO term of interest among the TGs in that regulon, "GO_enrich_term" the corresponding GO term, "GO_enrich_desc" the description of that GO term, and "#TGs_withGO" the number of TGs in the regulon that are associated with that GO term.

(e) **Regulon ranking** (Table 5): The column "borda_rank" shows the global Borda rank over all cluster-specific networks, while "borda_clusterRank" gives the Borda rank of the regulon within the network of one single-cell cluster (*see* **Note 16**). The Borda rank utilizes the cluster specificity, network centrality of the TF, and optionally also the

Table 5
Ranked regulons result table, columns with regulon ranking information

borda_rank	borda_clusterRank
1	1
2	1
3	1
4	2
5	2

enrichment of a regulon's TGs for GO terms of interest to produce a final ranking of the regulons based on the most relevant regulon importance metrics (*see* **Note 10**).

2. **Regulons file** (*_regulons.tsv): This file contains the final MINI-EX network, with each line representing a regulon. It is a tab-delimited file with the TF of that regulon as the first column, the regulon's single-cell cluster as the second column, and a comma-separated list of the regulon's target genes as the last column.

```
AT3G55770  Cluster_0   AT4G21910,AT2G41290,AT2G45430,
                       AT3G56070,AT2G43290,...
AT3G55770  Cluster_32  AT2G35860,AT3G49390,AT5G62960,
                       AT1G47570,AT1G77440,...
AT3G55770  Cluster_31  AT2G18450,AT2G30620,AT1G50430,
                       AT1G51650,AT5G49760,...
AT3G55770  Cluster_3   AT5G26780,AT2G45300,AT3G53260,
                       AT1G27450,AT5G43910,...
AT3G55770  Cluster_25  AT3G46130,AT5G26780,AT2G45300,
                       AT3G53260,AT1G43160,...
```

3. **Edge table file** ("*_edgeTable.tsv"): This file (Table 6) combines the most important information about the network as a tab-separated edge list. Each line represents an edge in the network, with the first column showing the gene ID of the TF, the second column the gene ID of the TG, the third the cluster ID, the fourth the global Borda rank of the regulon, the fifth the cluster-specific Borda rank, and the sixth the GRNBoost2 evidence score of that edge, which is a measure for the degree of coexpression.
4. **TF info file** ("*_TF_info.tsv"): This file (Table 7) shows detailed information about how TFs in the initial TF list are filtered during the different steps in MINI-EX. The TF info file

Table 6
Edge table result file

TF	TG	cluster	borda_rank	borda_clusterRank	weight
AT1G01010	AT2G32240	Cluster_11	4584	183	139859E+16
AT1G01010	AT4G36990	Cluster_11	4584	183	785487E+14
AT1G01010	AT1G23440	Cluster_11	4584	183	433944E+15
AT1G01010	AT1G73260	Cluster_11	4584	183	418308E+15
AT1G01010	AT4G01070	Cluster_11	4584	183	377707E+15

is a table with a TF on each row. The first column shows whether the TF is found as expressed in the count matrix, the second whether it is part of a regulon in the GRNBoost2 GRN, the third whether it is retained as a regulon after motif mapping filtering in **step 2** (Fig. 2), and the fourth if the TF is part of the final networks after expression filtering in **step 3**. In case the TF is expressed (included in the expression matrix), all following columns show the percentage of cells in which the TF is expressed for each single-cell cluster.

5. **GRNBoost2 network** ("*_grnboost2.tsv"): This file contains the GRN inferred by GRNBoost2. Each line in this file corresponds to a directed edge in this network, with the first column being the gene ID of a TF and the second column the gene ID of its TGs. The third column is the evidence score calculated by GRNBoost2 for this edge.

AT3G44750	AT4G12600	228.43688171026278
AT1G63100	AT1G50490	226.98249498720028
AT3G44750	AT5G27120	207.9133763122063
AT1G63100	AT3G51720	206.86956547679816
AT3G44750	AT4G25630	204.25228196977886

6. **GO enrichment of regulons** ("*_enricherGO.txt"): This file (Tables 8 and 9) contains the raw output of GO enrichment on the TGs of each regulon. The first column contains the gene set ID, denoting which set of genes (here the TGs of a regulon) was tested for enrichment. The set ID is constructed as the regulon's TF gene ID, followed by the cluster name of that regulon, joined by an underscore (e.g., AT3G44750_Cluster_4). The second column is the enriched GO term and the third is the p-value of the enrichment. The fourth column contains the q-value (adjusted p-value by using the Benjamini-Hochberg procedure [36]) and the fifth column

Table 7
Transcription factor information table

	isTF_expressed	isTF_expressionRegulons	isTF_TFBSRegulons	isTF_finalRegulons	LRC-0	LRC-1	...
AT1G01010	1	1	1	1	1.4	1.5	...
AT1G01030	1	1	0	0	2.3	0.5	...
AT1G01060	1	1	1	1	0.8	1.2	...
AT1G01250	0	0	0	0			...
AT1G01260	1	1	1	1	10.1	15.5	...

Table 8
Gene ontology enrichment output file (part 1)

#set_id	ftr_id	p-val	q-val	enr_fold
AT1G14900_Cluster_32	GO:0000003	0.000563632	0.00224408	2.56505
AT1G14900_Cluster_32	GO:0003006	0.00186918	0.00596871	2.45494
AT1G14900_Cluster_32	GO:0006139	0.00268205	0.0079436	2.08545
AT1G14900_Cluster_32	GO:0006351	0.000607938	0.00238854	2.64713
AT1G14900_Cluster_32	GO:0006355	0.000541584	0.00217134	2.67711

Table 9
Gene ontology enrichment output file (part 2)

set_size	ftr_size	ftr_ic	n_hits	hits
202	688	0.28924	16	AT5G61230,AT1G72150,AT4G30520,...
202	629	0.298994	14	AT1G72150,AT4G30520,AT2G32280,...
202	952	0.253911	18	AT3G61850,AT4G29230,AT5G60850,...
202	625	0.299688	15	AT3G61850,AT4G29230,AT5G60850,...
202	618	0.300913	15	AT3G61850,AT4G29230,AT5G60850,...

the enrichment fold (the ratio between the observed fraction of hits in the gene set and the fraction in the background). The sixth column shows the size of the gene set (i.e., number of TGs in the regulon), the seventh column shows the feature set size (i.e., the total number of genes associated with the GO term), and the eighth column shows the information content (IC) [37] of the GO term, which is a metric that ranges between 0 and 1 that describes how general (low values) or specific (high values) the GO term is. The ninth column shows the number of genes that the gene set and feature set have in common, and the last column displays those genes as a comma-separated list of gene IDs.

7. **Regmap figure** ("*_regmap_*.svg"): This figure (Fig. 3) is very informative to identify the most relevant TFs in different clusters. Several regmap figures are generated, each visualizing an increasing number (as indicated in the file name; e.g., *_regmap_25.svg) of the (e.g. 25) top most important regulators per cluster. The figure shows a grid in which the rows correspond with TFs and the columns with different single-cell clusters. Columns are also annotated by their cell type identity, as given in the cell cluster annotation input file, and if an optional index column is added to this file (see Subheading

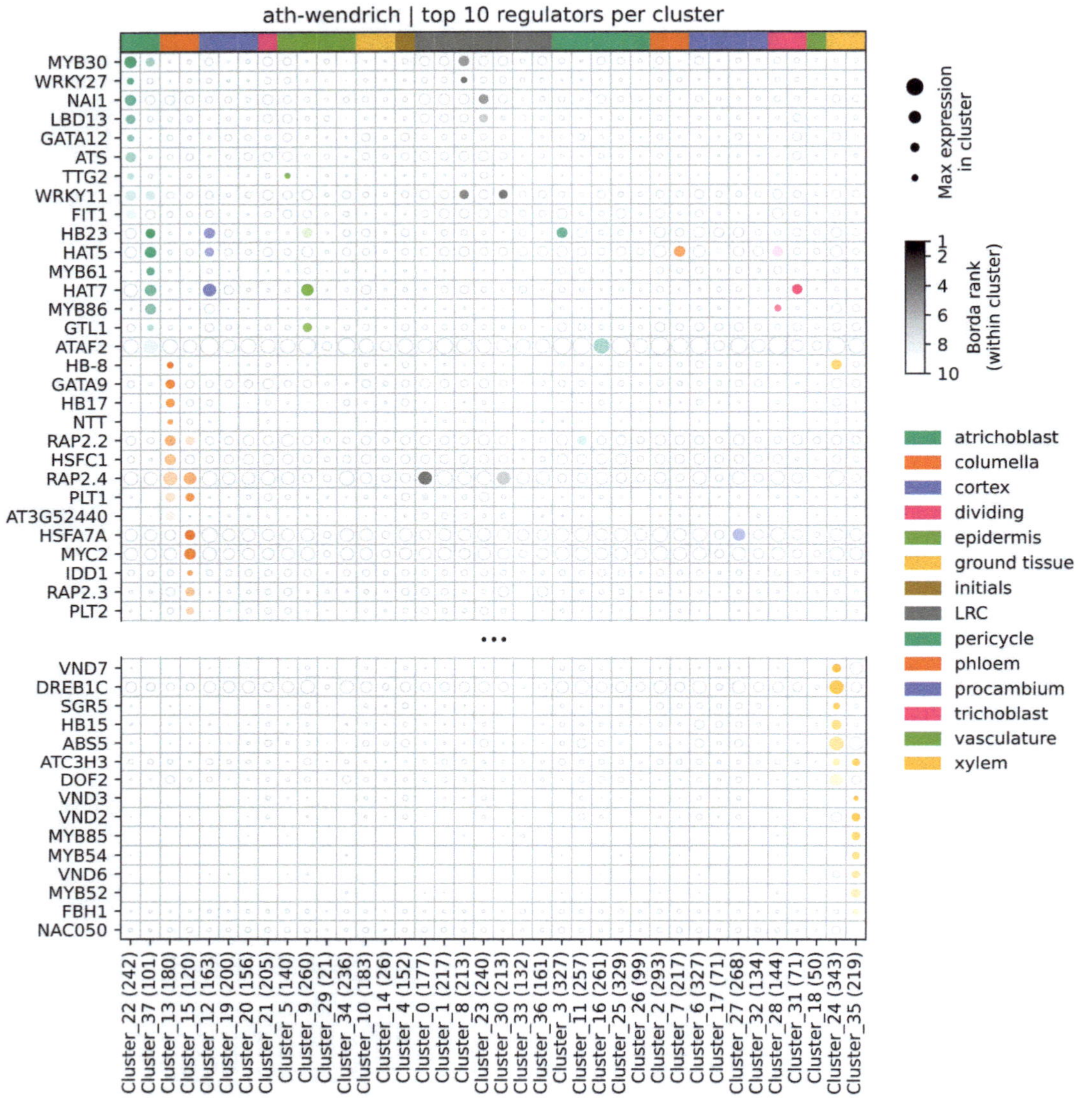

Fig. 3 Regmap figure of the Wendrich et al. [4] results. The top 10 most important regulators of each cluster (based on the cluster-specific Borda rank) are selected. Gene aliases are shown on the y-axis and clusters on the x-axis. For each cluster-specific network, the total number of regulons is added after the cluster name. The shade of the color of each circle corresponds to the cluster-specific Borda ranking of the given regulon, while the size corresponds to the maximum expression level of the regulator within that cluster

3.4), the specified column index is used in the plot. In each slot, a circle is plotted where the size of the circle corresponds to the maximum expression level (average of three highest expressing cells) of the TF in the given cluster and the shade of the color corresponds to the cluster-specific Borda rank. This plot allows one to visually identify the most important regulators in a given cell type, according to the Borda ranking procedure. One can also easily identify shared regulators, active in multiple cell types. If the clusters follow a known developmental trajectory, patterns of increasing or diminishing regulator importance can be tracked over that trajectory.

8. **Clustermap figure** ("*_clustermap.svg"): This figure (Fig. 4) shows a clustered heatmap of the regulon size in different clusters. The x-axis represents the different single-cell clusters and the y-axis the different TFs that can regulate different sets of target genes in different clusters. The heatmap shows the number of TGs in the regulon of the given cluster. This plot visualizes similarities between single-cell clusters based on their regulons.
9. **DE and specificity heatmap figures** ("*_heatmapDEcalls.svg" and "*_heatmapSpecificity.svg"): Both figures show the different single-cell clusters on the x-axis and the different TFs on the y-axis. The TFs are annotated with green if they are a relevant TF (i.e., they match a GO term of interest), with yellow if they are a known TF (annotated to any nonrelevant GO term), or with gray if they are not annotated to a GO term. For the DE heatmap (Fig. 5), the colors in the heatmap indicate whether the TF is upregulated (blue), expressed in at least 10% of the cells of that cluster (white), or expressed in less than 10% of the cells of that cluster (gray). For the specificity heatmap figure, the negative logarithm of the cluster specificity q-values (see cluster specificity column in ranked regulons file) is shown. Both files visualize where TFs are expressed in a cluster-specific manner.
10. **Log file** (*_log.txt): This file contains a summary of the input files and parameters used in a given MINI-EX run, as well as summary statistics on the initial data and the inferred GRNs. Number of regulons and median regulon size are given after each filtering step in MINI-EX, allowing to identify the impact of each filtering step. Finally, the log file shows the regulon importance metrics and ranking procedure (*see* **Notes 9** and **10**) used for the Borda ranking of regulons in **step 4** (Fig. 2).

3.9 Interpreting the Results

Finally, to illustrate how the MINI-EX results can be used to extract interesting biological knowledge, we will here look more in-depth at the results of two of the most informative output files: the ranked regulons file and the regmap figure. As an example, we focus on the well-studied xylem tissue, to see what information regarding its regulation is captured by MINI-EX.

First, we can visually inspect the regmap figure to identify the most important regulators. The xylem tissue consists of clusters 24 and 35, which correspond to the last two columns in the figure. At the lower part of the regmap figure, many known xylem differentiation regulators can be observed such as MYB52, MYB54,

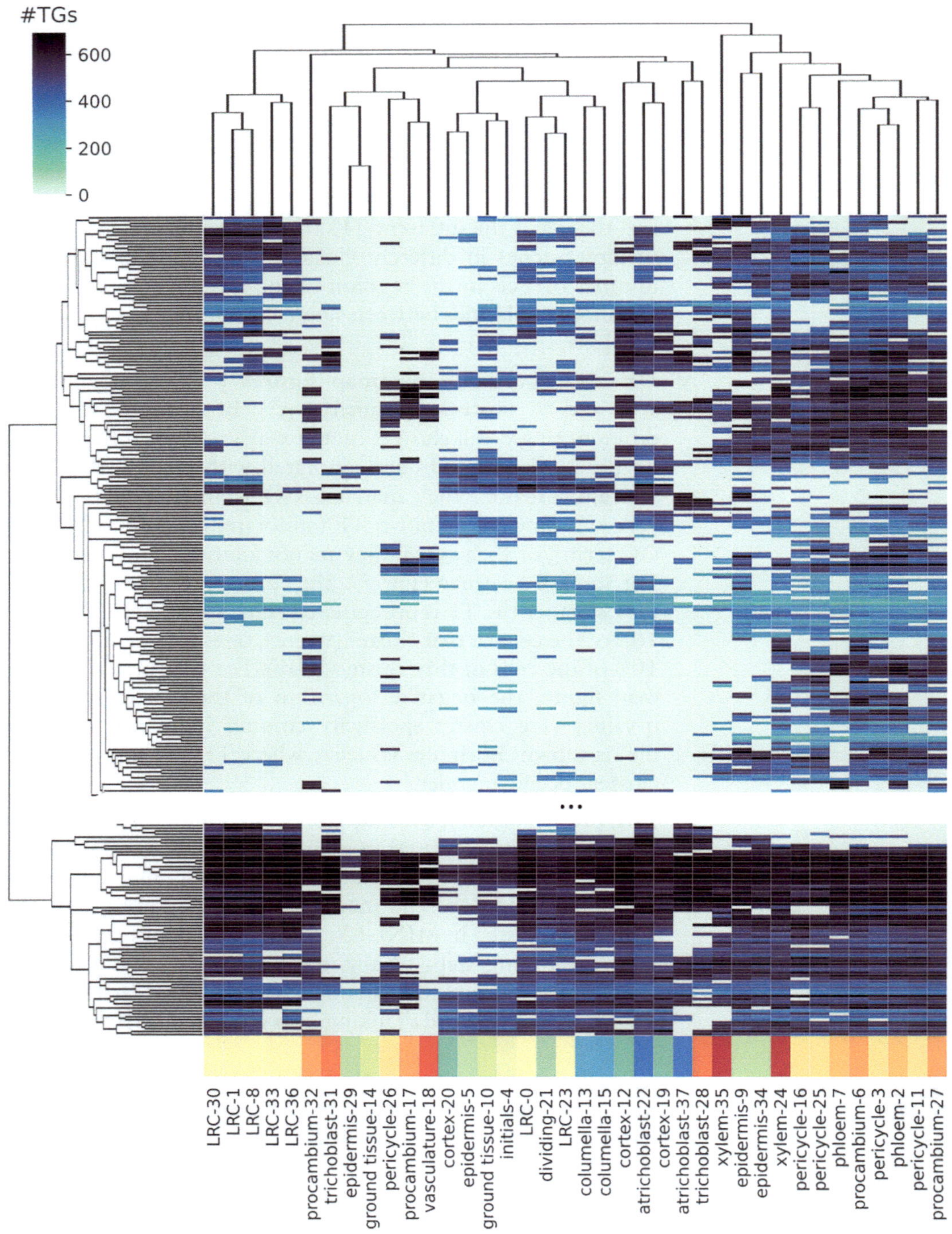

Fig. 4 Clustermap figure of the Wendrich et al. [4] results. This figure shows a clustered heatmap of the regulon size. The x-axis represents single-cell clusters and the y-axis individual regulons. The color of the heatmap represents the number of target genes the regulon has

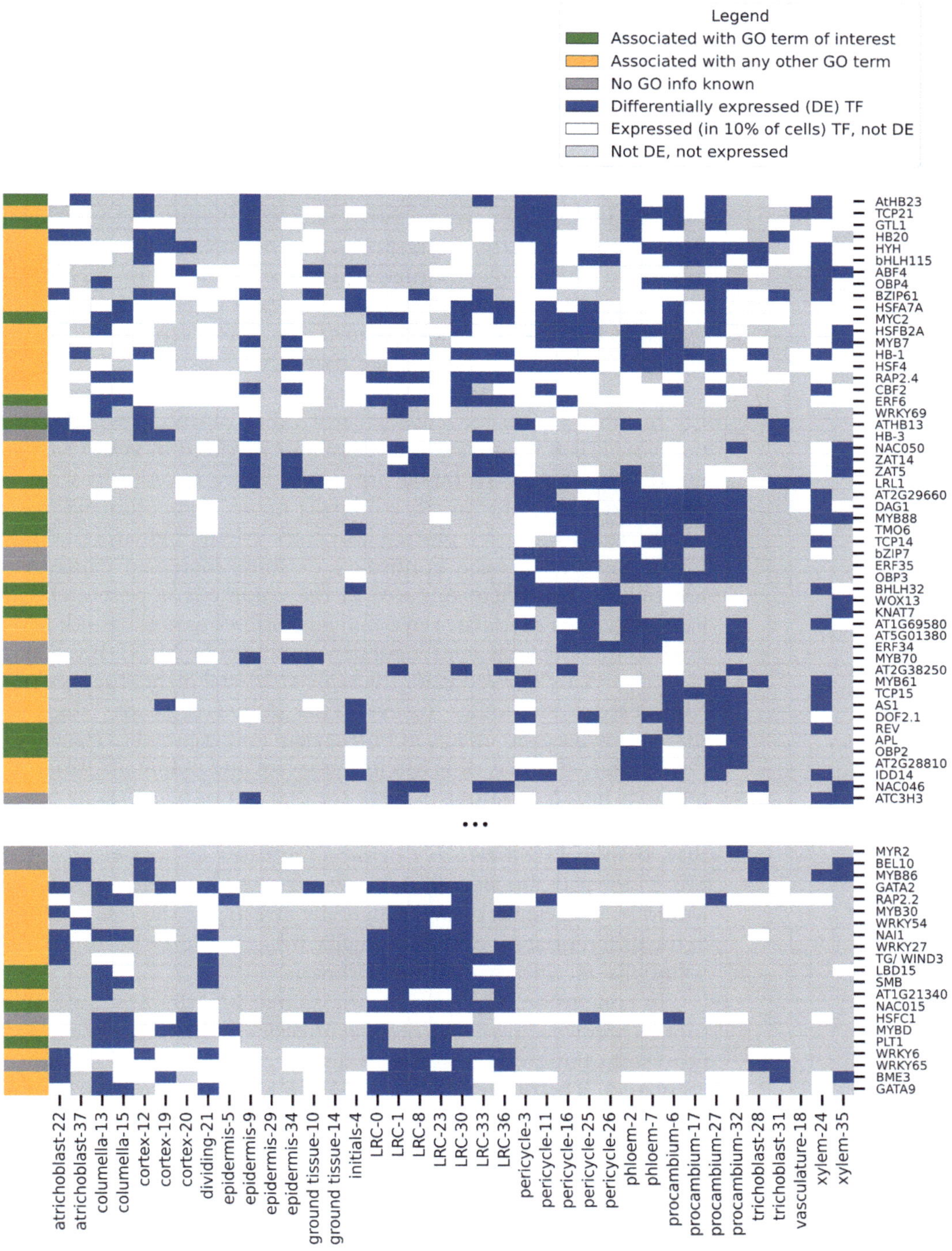

Fig. 5 DE heatmap figure of the Wendrich et al. [4] results. This figure shows a heatmap with the differentially expressed (DE) transcription factors (TFs) in at least one single-cell cluster. The x-axis represents single-cell clusters and the y-axis individual TFs. The color of the heatmap is blue if the TF is DE (= marker gene) in the cluster, white if it is not DE but it is expressed in more than 10% of cells, and gray if it is neither DE nor expressed in more than 10% of cells in the cluster. TFs are annotated as green if they are associated with a GO term of interest, yellow if they are associated with any other GO term, and gray if they are not associated with a GO term

MYB85, VND2, VND3, VND6, and VND7 [38]. The regmap shown in Fig. 3 is the smallest of several generated regmap figures, only coloring the top 10 best ranked regulators per cluster. While this regmap is the most readable, potential relevant regulators can be missed in this regmap, for example, the known xylem regulators MYB46 and MYB83, respectively, ranked 18th and 20th in xylem cluster 35. Similarly, correspondence between clusters can be missed in the smaller regmaps. This is the case for VND7 in Fig. 3, which is the top regulator for xylem cluster 24 but is not shown to be relevant for xylem cluster 35. However, looking at the ranked regulons file, we can see that VND7 is still ranked 13th, indicating a relevant role for this regulator in both clusters.

Another putative xylem regulator HB-8 can be seen higher on the plot. HB-8 is also a known regulator involved in xylem formation [39]. It is shown higher up in the figure because it is also a well-ranked regulator in the columella tissue, and columella regulators are (arbitrarily) plotted prior to xylem regulators. While HB-8 has indeed been confirmed to show increased expression not only in the xylem but also in the columella, a potential role for this gene as a regulator in columella differentiation has not been described, indicating a potential new discovery by MINI-EX. While both tissues are derived from the same stem-cell niche, they are not considered very similar tissues. This shared regulator suggests shared downstream effects in both xylem and columella. However, if we extract the TGs of HB-8 in the xylem and columella from the regulons file (*_regulons.tsv), we see that out of 672 TGs in xylem and 526 TGs in columella, only 27 TGs are shared. This suggests that, besides HB-8 possibly being a regulator in both tissues, the HB-8 downstream signaling is rewired. Based on these results, it would be interesting to experimentally investigate the role of HB-8 in the differentiation of the columella, for example, by performing a columella-specific knockout experiment.

In conclusion, this example shows that MINI-EX is a valuable tool to predict known and novel regulators in different tissues and processes of interest, as well as to discover new functions of known regulators. The interpretation of MINI-EX's summarizing output files and figures can offer researchers profound insights, enabling them to formulate new hypotheses about the biological function of regulators and their downstream effects.

4 Notes

1. In **step 2**, filtering is performed on the regulons of the initial regulatory network in two ways. First, a regulon is only kept if the TF motif is enriched in its TGs (i.e., more TGs than expected by chance have the TF binding motif). Second, for the remaining regulons, only TGs that have the TF motif are

retained. As a result, regulons under the control of a TF lacking information on their binding motif are removed. While MINI-EX by default extends motifs to all family members of that TF protein family (see "motif filter" parameter; Subheading 3.6), increasing the number of TFs with motif information, still a fraction of TFs (around 20% in *A. thaliana*) lacks this information which causes them to be filtered out during this step. Thus, even when extending motifs to the TF family, it is important to take into account that when little motif information is available, this can result in a filtering in **step 2** that is too severe. If one notices that too many relevant regulons are removed during **step 2** due to lack of motif information, it might be beneficial to omit **step 2** (see "do motif analysis" parameter; Subheading 3.6). However, for species (e.g., supported species) with sufficient TF motif information, it is recommended to include the motif mapping step as it has been shown to improve precision of the resulting GRNs [19, 40, 41].

2. In MINI-EX version 2.0, the way in which enrichment statistics were calculated was modified for all enrichment steps within MINI-EX (**steps 2**, **3a**, and **4a** in Fig. 2). A first modification is to perform enrichment using a background model of only expressed genes [27, 28]. This is implemented by first inferring the set of expressed genes based on the count matrix given as input. Then, the motif mapping file, the marker gene file, and the GO file, used in the three enrichment steps of MINI-EX, are filtered to only keep expressed genes. Lastly, the filtered files are used and the set of all expressed genes is used as a background model when calculating enrichment statistics. A second modification was also made so that no enrichment statistics were calculated if there is only one hit in the gene set.

3. MINI-EX provides gene-GO association files for supported species, derived from PLAZA [29] and combined with GO information from TAIR (www.arabidopsis.org) [42] for *A. thaliana*. Originally, in MINI-EX, these contained only GO terms with experimental evidence codes (EXP, IMP, IDA, IPI, IGI, IEP, TAS, NAS, IC). Since version 2.1, all evidence codes including computational evidence codes are included for all supported species except *A. thaliana*, in order to cover a sufficient number of genes to calculate GO enrichment statistics. Given that for the model species *A. thaliana*, sufficient experimental GO information is available, a more stringent selection for high-quality (i.e., experimental) terms is opted for (since version 2.2, this also includes the high-throughput experiment code HTP, but not the non-traceable author statement code NAS).

4. The expression matrix provided to GRNBoost2 as input should be a count matrix. Due to the nonlinear tree-based approach

GRNBoost2 takes, normalization of this matrix is possible, but not necessary [43]. It is important to note that normalization procedures such as log-transformation or transcripts per million can introduce artifacts and should therefore not be applied [44]. One possible normalization method is SCT normalization, as implemented by the R package Seurat. With this method, Pearson residuals are reverse-transformed into a count matrix that is adjusted for sequencing depth and corresponds to the "counts" matrix of the SCT slot in a Seurat object. Such an adjusted count matrix can be provided to MINI-EX as input. While such normalization procedures can help to reduce some biases, they are not necessary. Therefore, we recommend using unnormalized, raw counts to new users as a valid first option.

5. In **step 3a** (Fig. 2), MINI-EX uses a set of differentially expressed genes (DEGs; i.e., marker genes) per cluster. These DEGs should be upregulated genes (set by specifying the "only.pos = TRUE" parameter in the Seurat function FindAllMarkers). MINI-EX, as it is currently implemented (version 2.2), is only intended to infer upregulatory relationships between TF and TG. Hence, only upregulated DEGs should be given as input.
6. The motif mapping file was generated by mapping known TF binding site (TFBS) motifs to the regulatory region around all genes in the genome. If the TFBS motif of a TF is found around the gene, the motif-TG pair is added to the motif mapping file. As the aim of this chapter is to be a practical guide on how to use MINI-EX, we will not give an extensive description on the procedures used to generate the precomputed motif mapping files of which MINI-EX makes use. More detail regarding the way this file was created can be found in [19, 31, 32].
7. The default parameters of MINI-EX are based on benchmarking results on *A. thaliana* described in the original MINI-EX publication. We recommend evaluating how well these parameters perform, especially when using MINI-EX for species other than *A. thaliana*, using a relevant gold standard reporting TF-TG interactions or a set of known TFs for the process of interest [19].
8. The terms of interest file contains words that can be linked to specific GO terms of interest. Using these words, MINI-EX retrieves the GO terms that contain at least one of the provided words in their description. For example, adding the term "trichome" in the terms of interest file will result in several GO terms being selected such as "trichome branching," "regulation of trichome morphogenesis," and others. We encourage users to provide synonyms and related terms such as "xylem,"

"vessel," and "tracheary" in order to cover all relevant GO terms. Also, when one has a broad research question such as interest in all root related processes, we suggest providing not only "root" as a term but also more specific tissues or processes such as "trichoblast," "columella," etc.

9. When no terms of interest file is provided or no GO information is available, MINI-EX uses its default ranking procedure, also referred to as the standard or "std" procedure. This procedure ranks all inferred regulons based on four regulon importance metrics. These are the out-degree, closeness, betweenness (three network centrality metrics), and cluster specificity. Cluster specificity corresponds with the q-value of enrichment between cluster-specific marker genes and the TGs of a regulon (i.e., how specific are the TGs of the regulon for the cluster) calculated during **step 3a** (Fig. 2). These four regulon importance metrics are integrated using unweighted Borda counts. Here, the regulons are ranked on each of the metrics and the geometric mean of those four ranks is taken for each regulon. The resulting number is used to make the final ranking.
10. When both GO information is provided and the user specified a set of terms of interest, MINI-EX will use this information as reference to tailor its ranking procedure to the research question of the user, defined by these specific terms. This ranking procedure is referred to as the reference of "ref" procedure. It uses the same regulon importance metrics as the standard procedure (out-degree, closeness, betweenness, and cluster specificity), with the addition of GO enrichment q-value between the TGs of a regulon and GO terms of interest (*see* **Note 3** and **8**). The reference procedure makes use of weighted Borda counts to integrate regulon importance metrics and automatically selects which metrics to include for this. Giving weights and selecting regulon importance metrics is done based on R50 values. The R50 value is the rank at which 50% of all relevant regulons (i.e., regulons with a TF that is associated with a GO term of interest) are retrieved. Based on the ranking computed on each individual metric, an R50 value is calculated for each metric. Then, ranks for each metric are divided by the R50 of that metric and subsequently summed to get the weighted Borda score and final ranking. As such, metrics that perform worse at ranking relevant TFs and thus have a high R50 will contribute less to the final ranking. Finally, not all five regulon importance metrics are necessarily used to calculate the final Borda rank. MINI-EX will calculate a weighted Borda rank as described above for each possible combination of at least one regulon importance metric and select the optimal combination of metrics as the one that

results in weighted Borda ranks with the best R50 value, which are used as the final ranking in the MINI-EX output. The selected combination of metrics can be found in the MINI-EX log file (since version 2.2).

11. The use of the TF family motif extension strategy aims to increase the recall of the results and is based on the observation that the binding motifs of different members of a TF protein family are often very similar due to their shared ancestry. Thus, to prevent filtering out many regulons only due to their TFs lacking motif information, this approach uses the known binding motifs of all family members as a proxy, thereby covering more TFs. However, while this strategy has been shown to increase recall, it may reduce precision because when TFs with motif information available are expanded to the family level, more spurious matches can be found that introduce noise, giving rise to more false positive TF-TG interactions in the final GRNs.

12. In step 3a, cluster-specific marker genes (i.e., differentially expressed genes calculated for each single-cell cluster) are used for enrichment and filtering (Fig. 2). In this step, regulons with TGs that are not enriched for the top N (the "top markers" parameter) most significant marker genes for each cluster are filtered out. For the remaining regulons, only TGs that are in the cluster-specific marker genes set are retained. Therefore, changing the "top markers" parameter strongly affects overall regulon sizes.

13. The "expression filter" parameter corresponds to the percentage of cells in a cluster that has to express a TF, in order for that TF to be considered expressed in that cluster. This metric does not take into account the expression level of a gene, which means that if the TF has one UMI count in a cell, the TF is considered expressed in that cell. As a result, the use of count normalization methods that can impute counts when the raw count was zero can result in a too liberal filtering in **step 3b**. Conversely, on datasets with relatively low coverage, the high number of zero counts can result in too stringent filtering when using the default. Thus, as the optimal value for this parameter depends on your dataset and preprocessing steps, it is recommended to assess how many regulons are filtered out during this step (by looking at the log file and "TF info file") and adjust this parameter accordingly. An example of such adjustment can be found in Ferrari et al. (2022) [19] when running MINI-EX on maize.

14. If the user attempts to execute MINI-EX on an HPC cluster with an executor other than SLURM without modifying the "name" parameter within the "executor" scope, MINI-EX will

result in failure with the following error message: "cannot run program 'sbatch'" (where "sbatch" is the command used by SLURM to submit jobs).

15. In MINI-EX, three network centrality metrics are calculated (out-degree, betweenness, and closeness), as implemented by the python library NetworkX (version 2.5.1) [45]. The out-degree centrality corresponds to the number of outgoing edges of a TF (i.e., the number of TGs), divided by the largest possible out-degree, which is the number of other nodes in the network of that cluster. The betweenness of a TF is the fraction of shortest paths between any two nodes (excluding the TF) containing that TF. Closeness is the number of all other reachable nodes from a TF, divided by the average path length of a TF to all other nodes in the network. These metrics are complementary ways to estimate the importance of a TF in the network structure. Note that each of these metrics is calculated for the cluster-specific GRN the regulon is part of, independent of the other cluster-specific GRNs.
16. The Borda rank is calculated both in a global way and in a cluster-specific manner. For the global ranking, the regulons of all cluster-specific networks are taken into account when obtaining the ranks based on the different regulon importance metrics (which are themselves calculated within the cluster-specific GRN of that regulon; *see* **Note 15**), often including regulons under the control of the same TF, but in a different single-cell cluster with different TGs. This is opposed to the cluster-specific ranking that makes use of the same regulon importance metric values as the global ranking but only creates a ranking of regulons within one cluster. Based on the ranks computed on each individual metric, the Borda rank is calculated in the same way for both the global and cluster-specific ranking.

Acknowledgments

This work was supported by an Industrieel Onderzoeksfonds grant from Ghent University (F2020/IOF-StarTT/151; IOF. PRO.2021.0017) to S. L. and K. V. and a Bijzonder Ondersoeksfonds grant from Ghent University (BOF24Y2019001901) to N. M. and the Research Foundation-Flanders (FWO; Odysseus II G0D0515N) to J.S. We thank Herman De Beukelaer for helping us to improve enrichment analysis and Camilla Ferrari for developing the initial version of MINI-EX.

Conflict of Interest The authors have no conflicts of interest to declare that are relevant to the content of this chapter.

References

1. Nolan TM, Shahan R (2023) Resolving plant development in space and time with single-cell genomics. Curr Opin Plant Biol 76:102444. https://doi.org/10.1016/j.pbi.2023.102444
2. Otero S, Gildea I, Roszak P et al (2022) A root phloem pole cell atlas reveals common transcriptional states in protophloem-adjacent cells. Nat Plants 8:954–970. https://doi.org/10.1038/s41477-022-01178-y
3. Tung C-C, Kuo S-C, Yang C-L et al (2023) Single-cell transcriptomics unveils xylem cell development and evolution. Genome Biol 24: 3. https://doi.org/10.1186/s13059-022-02845-1
4. Wendrich JR, Yang B, Vandamme N et al (2020) Vascular transcription factors guide plant epidermal responses to limiting phosphate conditions. Science 370. https://doi.org/10.1126/science.aay4970
5. Seyfferth C, Renema J, Wendrich JR et al (2021) Advances and opportunities in single-cell transcriptomics for plant research. Annu Rev Plant Biol 72:847–866. https://doi.org/10.1146/annurev-arplant-081720-010120
6. Denyer T et al (2019) Spatiotemporal developmental trajectories in the Arabidopsis root revealed using high-throughput single-cell RNA sequencing. Dev Cell 48:840–852.e845
7. Jean-Baptiste K, McFaline-Figueroa JL, Alexandre CM et al (2019) Dynamics of gene expression in single root cells of arabidopsis thaliana. Plant Cell 31:993–1011. https://doi.org/10.1105/tpc.18.00785
8. Ryu KH, Huang L, Kang HM, Schiefelbein J (2019) Single-cell RNA sequencing resolves molecular relationships among individual plant cells. Plant Physiol 179:1444–1456. https://doi.org/10.1104/pp.18.01482
9. Shulse CN, Cole BJ, Ciobanu D et al (2019) High-throughput single-cell transcriptome profiling of plant cell types. Cell Rep 27: 2241–2247.e4. https://doi.org/10.1016/j.celrep.2019.04.054
10. Zhang T-Q, Xu Z-G, Shang G-D et al (2019) A single-cell RNA sequencing profiles the developmental landscape of arabidopsis root. Mol Plant. https://doi.org/10.1016/j.molp.2019.04.004
11. Liu Q, Liang Z, Feng D et al (2021) Transcriptional landscape of rice roots at the single-cell resolution. Mol Plant 14:384–394. https://doi.org/10.1016/j.molp.2020.12.014
12. Marand AP, Chen Z, Gallavotti A, Schmitz RJ (2021) A cis-regulatory atlas in maize at single-cell resolution. Cell 184:3041–3055.e21. https://doi.org/10.1016/j.cell.2021.04.014
13. Ortiz-Ramírez C, Guillotin B, Xu X et al (2021) Ground tissue circuitry regulates organ complexity in maize and Setaria. Science 374:1247–1252. https://doi.org/10.1126/science.abj2327
14. Omary M, Gil-Yarom N, Yahav C et al (2022) A conserved superlocus regulates above- and belowground root initiation. Science 375: eabf4368. https://doi.org/10.1126/science.abf4368
15. De Clercq I, Van de Velde J, Luo X et al (2021) Integrative inference of transcriptional networks in Arabidopsis yields novel ROS signalling regulators. Nat Plants 7:500–513. https://doi.org/10.1038/s41477-021-00894-1
16. Depuydt T, De Rybel B, Vandepoele K (2023) Charting plant gene functions in the multi-omics and single-cell era. Trends Plant Sci 28: 283–296. https://doi.org/10.1016/j.tplants.2022.09.008
17. Wu T-Y, Goh H, Azodi CB et al (2021) Evolutionarily conserved hierarchical gene regulatory networks for plant salt stress response. Nat Plants 7:787–799. https://doi.org/10.1038/s41477-021-00929-7
18. Kulkarni SR, Vandepoele K (2020) Inference of plant gene regulatory networks using data-driven methods: a practical overview. Biochim Biophys Acta Gene Regul Mech 1863:194447
19. Ferrari C, Pérez NM, Vandepoele K (2022) MINI-EX: integrative inference of single-cell gene regulatory networks in plants. Mol Plant 15:1807–1824. https://doi.org/10.1016/J.MOLP.2022.10.016
20. Cao S, He Z, Chen R et al (2023) scPlant: a versatile framework for single-cell transcriptomic data analysis in plants. Plant Commun 4:100631. https://doi.org/10.1016/j.xplc.2023.100631
21. Guillotin B, Rahni R, Passalacqua M et al (2023) A pan-grass transcriptome reveals patterns of cellular divergence in crops. Nature 617:785–791. https://doi.org/10.1038/s41586-023-06053-0
22. Di Tommaso P, Chatzou M, Floden EW et al (2017) Nextflow enables reproducible computational workflows. Nat Biotechnol 35:316–319
23. Kurtzer GM, Sochat V, Bauer MW (2017) Singularity: scientific containers for mobility of compute. PLoS One 12:e0177459. https://doi.org/10.1371/journal.pone.0177459

24. Ke Y, Minne M, Eekhout T, De Rybel B (2023) Single cell RNA-sequencing in arabidopsis root tissues. Methods Mol Biol 2698: 41–56. https://doi.org/10.1007/978-1-0716-3354-0_4
25. Hafemeister C, Satija R (2019) Normalization and variance stabilization of single-cell RNA--seq data using regularized negative binomial regression. Genome Biol 20:296. https://doi.org/10.1186/s13059-019-1874-1
26. Moerman T, Aibar Santos S, Bravo González-Blas C et al (2019) GRNBoost2 and Arboreto: efficient and scalable inference of gene regulatory networks. Bioinformatics 35:2159–2161. https://doi.org/10.1093/bioinformatics/bty916
27. Rhee SY, Wood V, Dolinski K, Draghici S (2008) Use and misuse of the gene ontology annotations. Nat Rev Genet 9:509–515. https://doi.org/10.1038/nrg2363
28. Timmons JA, Szkop KJ, Gallagher IJ (2015) Multiple sources of bias confound functional enrichment analysis of global -omics data. Genome Biol 16:186. https://doi.org/10.1186/s13059-015-0761-7
29. Van Bel M, Silvestri F, Weitz EM et al (2022) PLAZA 5.0: extending the scope and power of comparative and functional genomics in plants. Nucleic Acids Res 50:D1468–D1474. https://doi.org/10.1093/nar/gkab1024
30. Jin J, Tian F, Yang D-C et al (2017) PlantTFDB 4.0: toward a central hub for transcription factors and regulatory interactions in plants. Nucleic Acids Res 45:D1040–D1045. https://doi.org/10.1093/nar/gkw982
31. Manosalva Pérez N, Vandepoele K (2023) Prediction of transcription factor regulators and gene regulatory networks in tomato using binding site information. Methods Mol Biol 2698:323–349. https://doi.org/10.1007/978-1-0716-3354-0_19
32. Manosalva Pérez N, Ferrari C, Engelhorn J et al (2024) MINI-AC: inference of plant gene regulatory networks using bulk or single-cell accessible chromatin profiles. Plant J 117:280–301. https://doi.org/10.1111/tpj.16483
33. Castro-Mondragon JA, Riudavets-Puig R, Rauluseviciute I et al (2022) JASPAR 2022: the 9th release of the open-access database of transcription factor binding profiles. Nucleic Acids Res 50:D165–D173. https://doi.org/10.1093/NAR/GKAB1113
34. Weirauch MT, Yang A, Albu M et al (2014) Determination and inference of eukaryotic transcription factor sequence specificity. Cell 158:1431–1443. https://doi.org/10.1016/j.cell.2014.08.009
35. Klopfenstein DV, Zhang L, Pedersen BS et al (2018) GOATOOLS: a python library for gene ontology analyses. Sci Rep 8:10872. https://doi.org/10.1038/s41598-018-28948-z
36. Benjamini Y, Hochberg Y (1995) Controlling the false discovery rate - a practical and powerful approach to multiple testing. J R Stat Soc Ser B-Methodol 57:289–300
37. Resnik P (2011) Semantic similarity in a taxonomy: an information-based measure and its application to problems of ambiguity in natural language. CoRR abs/1105.5
38. Hussey SG, Mizrachi E, Creux NM, Myburg AA (2013) Navigating the transcriptional roadmap regulating plant secondary cell wall deposition. Front Plant Sci 4:325. https://doi.org/10.3389/fpls.2013.00325
39. Baima S, Nobili F, Sessa G et al (1995) The expression of the Athb-8 homeobox gene is restricted to provascular cells in Arabidopsis thaliana. Development 121:4171–4182. https://doi.org/10.1242/dev.121.12.4171
40. McCalla SG, Fotuhi Siahpirani A, Li J et al (2023) Identifying strengths and weaknesses of methods for computational network inference from single-cell RNA-seq data. G3 (Bethesda) 13:jkad004. https://doi.org/10.1093/g3journal/jkad004
41. Siahpirani AF, Roy S (2017) A prior-based integrative framework for functional transcriptional regulatory network inference. Nucleic Acids Res 45:e21. https://doi.org/10.1093/nar/gkw963
42. Berardini TZ, Reiser L, Li D et al (2015) The Arabidopsis information resource: making and mining the "gold standard" annotated reference plant genome. Genesis 53:474–485. https://doi.org/10.1002/dvg.22877
43. Aibar S, González-Blas CB, Moerman T et al (2017) SCENIC: single-cell regulatory network inference and clustering. Nat Methods 14:1083–1086. https://doi.org/10.1038/nmeth.4463
44. Crow M, Paul A, Ballouz S et al (2016) Exploiting single-cell expression to characterize co-expression replicability. Genome Biol 17:101. https://doi.org/10.1186/s13059-016-0964-6
45. Hagberg AA, Schult DA, Swart P, Hagberg JM (2008) Exploring network structure, dynamics, and function using NetworkX. Proc Python Sci Conf

Chapter 13

A Practical Guide to Gene Regulatory Networks in Plants: From RNA Sequencing to Identification of Transcription Factor Binding Sites

Joaquín Grau and José M. Franco-Zorrilla

Abstract

Plant adaptive responses to their environment are orchestrated by transcriptional reprogramming, regulated by sequence-specific transcription factors (TFs) recognizing specific TF-binding sites (TFBSs), and understanding their regulatory grammar is crucial for unravelling plant adaptation mechanisms. Empirical data have contributed to the elucidation of TFBS sequence motifs involved in biological processes, but comprehensive experimental approaches may not be feasible, especially in non-model species. Different computational algorithms have facilitated the elucidation of TFBS sequence motifs and the construction of predictive models, shedding light on plant gene regulatory networks. In this context, the development of straightforward computational pipelines and easy-to-use bioinformatics tools is of particular relevance to make gene expression analysis accessible to the research community. This chapter presents a methodology to infer TF regulators applicable to 60 plant species from RNA sequencing (RNA-seq) data as the starting point. It includes RNA-seq analysis and quantification, gene clustering using WGCNA to identify co-regulated gene modules, and searching for enriched TFBSs associated with these modules. The methodology is illustrated using Arabidopsis RNA-seq data related to abiotic stress. By providing a user-friendly pipeline, researchers are empowered to unravel the molecular basis of gene expression dynamics in plants.

Key words Transcription factor (TF), DNA-binding sites, RNA-seq, WGCNA clustering, TFBS enrichment

1 Introduction

Plant adaptive responses to the environment most often involve a transcriptional reprogramming, which is tightly controlled by sequence-specific transcription factors (TFs). TFs are responsible for controlling the expression of target genes by recognizing specific DNA sequences, referred to as TF-binding sites (TFBSs). Transcriptional regulation is quite a complex process that involves the activity of hundreds or even thousands of TFs that can act either alone or cooperatively to regulate the expression of a multitude of genes. Therefore, transcriptional regulation is a combinatorial

José M. Franco-Zorrilla and Carmen Martín-Pizarro (eds.), *Plant Transcription Factors: Methods and Protocols*, Methods in Molecular Biology, vol. 2985, https://doi.org/10.1007/978-1-0716-4972-5_13,

result of *cis*- (i.e., TFBS in regulatory regions) and *trans*-acting, the TFs, elements. In this scenario, unraveling the regulatory grammar governing gene expression and discriminating TFBSs within the vast expanse of genomic sequences becomes a critical step in discerning the molecular basis of plant adaptation to the environment. The use of empirical data and computational algorithms has been essential to develop methodologies to elucidate the sequence motifs recognized by TFs and to elaborate predictive models that provide valuable insights into the regulatory networks that orchestrate plant biology.

Several in vivo and in vitro experimental approaches have contributed to describe hundreds of TFBS in plants [1]. Among in vivo strategies, chromatin immunoprecipitation and sequencing (ChIP-seq), DNase-seq, and Transposase-Accessible Chromatin and sequencing (ATAC-seq) [2–4] have been utilized to identify target genes or accessible chromatin regions, but direct identification of the TFBS has been more limited. In vitro approaches, such as universal Protein Binding Microarrays or DNA Affinity Purification and sequencing (DAP-seq), have allowed the identification of hundreds of plant TFBS and a clear visualization of a landscape of virtually all the plant TF families [5–7].

Curated TF-binding models are collected in several databases, such as Cis-BP, Cistrome, or JASPAR, for access to research community [5, 7, 8]. Some bioinformatic tools use the available information on TF-binding models to provide a stepwise analysis of custom lists of genes to predict TF-target genes and TFBS enrichment. Whereas some of these tools only cover *Arabidopsis* or just a small number of plant species, some other may require additional steps from the final user that are not always obvious for nonspecialists. With this regard, the web portal TDTHub offers the possibility of predicting enriched TFBS from lists of genes from 60 plant species in a few quick steps [9].

In this chapter, we present a comprehensive methodology to infer relevant TF regulators applicable for 60 plant species—including the most relevant crops—from lists of genes generated either from custom or publicly available RNA-seq data. The methodology here described represents a highly accessible pipeline to fulfill with all the computational steps in an easy manner with some minimal software installation and computational skills. The complete pipeline consist on the following steps:

(a) Analysis of RNA-seq samples and quantification of transcript levels from own or downloaded from repositories RNA-seq data.

(b) Gene clustering using Weighted Correlation Network Analysis (WGCNA) to identify gene modules highly correlated and, therefore, expected to share some common regulatory mechanisms. Gene clustering can be performed from complete gene sets or from prefiltered differentially expressed gene (DEG) sets.

(c) Search for enriched TFBS associated to the identified clusters of co-regulated gene sets to decipher the regulatory logic governing the observed transcriptional dynamics.

The pipeline is illustrated with some results generated after the analysis of 194 *Arabidopsis* RNA-seq samples corresponding to 56 bioprojects related with abiotic stress. By following this methodology, we empower researchers with a robust framework for deciphering the molecular basis of gene expression dynamics.

2 Materials

2.1 Computational Hardware

Personal computer with Internet connection, minimum requirement 16 Gb RAM (*see* **Note 1**).

2.2 Computational Software

1. Operating system: Linux Ubuntu 18.04 LTS or higher, Windows 10 or higher or Mac OS (*see* **Note 2**).
2. Updated Linux packages (*see* **Note 3**).
3. Miniconda3 with the specific channels priority setting for Nextflow (*see* **Note 4**).
4. Nextflow and nf-core installed in a conda virtual environment (*see* **Note 5**).

2.3 Sequence Data Files

1. Genome sequence (FASTA format, suffix ".fa" or ".fasta") and annotation (GFF format, suffix ".gff" or ".gff3") files (*see* **Note 6**).
2. RNA-seq data in FASTQ format and gzip-compressed (suffix ".fq.gz" or ".fastq.gz") (*see* **Note 7**).

3 Methods

The computational workflow detailed in this section consists of four main parts, as detailed in Fig. 1. The first part, RNA-seq quantification, includes all the sequential steps to quantify the gene expression levels starting from raw sequencing data. The second part is an optional module to identify differentially expressed genes between pairs of contrasts. The levels of expression are then subjected to a clustering analysis using Weighted Correlation Network Analysis (WGCNA). Finally, genes corresponding to each cluster are used for the identification of enriched TF-binding sites that could help to determine the most relevant transcriptional regulators involved in the biological process under study. Importantly, the entire workflow has been designed to perform each step

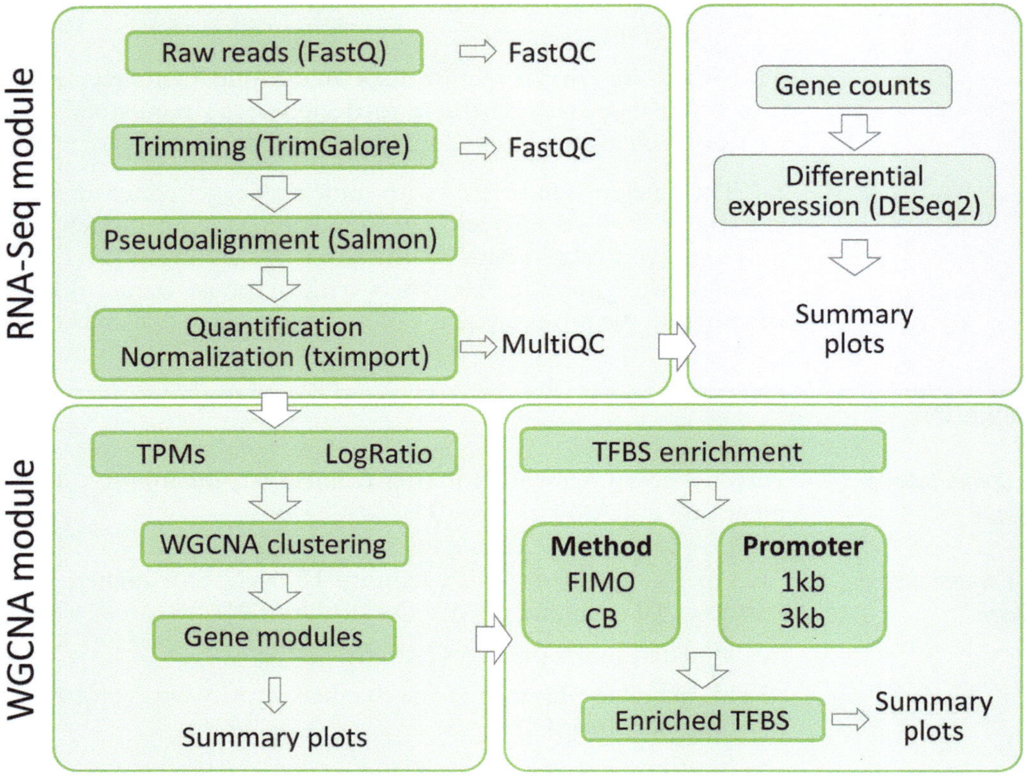

Fig. 1 Schematic overview of the pipeline to identify enriched TF-binding sites starting for RNA-seq data. The complete pipeline consists of three steps: RNA-seq module, to obtain the gene expression matrix; WGCNA module, to perform gene clustering using weighted gene correlation network analysis; and TDTHub, to obtain enriched TFBS from gene lists. The basic pipeline may be complemented with the analysis of differential expression for DEG-filtering (DE module)

sequentially with minimal intervention from the researcher by using optimized workflows (*see* **Note 8**).

3.1 RNA-Seq Quantification Module

This part will be performed following the "nf-core/rnaseq" pipeline previously installed (*see* **Note 5**). This pipeline maps sequencing data to an annotated reference genome and automatically returns quantified gene values. The sequential steps involve (i) quality control (QC) with FASTQC; (ii) trimming and quality filtering of raw reads using TrimGalore! [10]; (iii) quantification of transcript abundances carried out with the pseudoaligner Salmon [11]; (iv) transformation of transcript to gene expression values with the R library "tximport" [12]; and (v) generation of the gene expression matrix (Fig. 1). For illustration purposes, we are going to show a step-by-step analysis of an RNA-seq dataset obtained from SRA database (accession SRP058527), comprising control and two experimental conditions, with two replicates each.

3.1.1 Input Data Files

1. Create a folder (named "`wgcnamodules_testdata`" in this example) that will contain input files and where outputs of this pipeline will be stored. This can be done in the Linux console by typing.

```
mkdir wgcnamodules_testdata
cd wgcnamodules_testdata
```

2. Copy reference genome (FASTA), annotation (GFF), and FASTQ gzip-compressed files in the folder. This can be easily done from the console by using the command "`wget`" followed by the link to the file. In this example, we can obtain the *Arabidopsis thaliana* reference files from release 58 of Ensembl Plants by typing.

```
wget https://ftp.ebi.ac.uk/ensemblgenomes/pub/release-58/
plants/fasta/arabidopsis_thaliana/dna/Arabidopsis_thaliana.
TAIR10.dna.toplevel.fa.gz
wget https://ftp.ebi.ac.uk/ensemblgenomes/pub/release-58/
plants/gff3/arabidopsis_thaliana/Arabidopsis_thaliana.
TAIR10.58.gff3.gz
```

3. Copy FASTQ gzip-compressed files to a new folder named "`fastq`." Note that PE sequencing data involves two files per sample (*see* **Note 7**).
4. Create a metadata samplesheet specifying all the information of the samples in a CSV file and save in the "`fastq`" folder (*see* **Note 9**).

3.1.2 RNA-Seq Quantification

Once the input files are ready, executing a single command will run all the mentioned steps of the pipeline in a stepwise manner.

1. Replace folder and annotation file names by your own data names in the command below and run (*see* **Note 10**):

```
nextflow run nf-core/rnaseq -r 3.14.0 -profile conda --input
fastq/metadata_rnaseq.csv --outdir rna_seq --fasta Arabidop-
sis_thaliana.TAIR10.dna.toplevel.fa.gz --gff Arabidop-
sis_thaliana.TAIR10.58.gff3.gz --pseudo_aligner --
extra_salmon_quant_args '--seqBias --gcBias' --skip_alignment
true --max_cpus 6 --max_memory 8.GB
```

2. When it is finished, a folder for each executed result is generated. Open the file in "`wgcnamodules_testdata/rna_seq/multiqc/multiqc_report.html`" to find a complete report for sample QC, alignment scores, and other useful metrics for fully interpretation of RNA-seq data and quantification (Fig. 2).

A

General Statistics

Copy table | Configure Columns | Plot | Showing 6/6 rows and 6/18 columns.

Sample Name	% Aligned	M Aligned	% Dups	% GC	M Seqs	% BP Trimmed
CONTROL1_REP1	93.3%	8.1	47.6%	45%	9.2	6.4%
CONTROL1_REP2	95.0%	33.1	68.9%	45%	40.1	13.7%
TREATMENT1_REP1	91.9%	12.1	60.6%	45%	15.2	13.7%
TREATMENT1_REP2	89.7%	52.3	75.0%	46%	58.5	0.9%
TREATMENT2_REP1	94.4%	13.2	49.2%	46%	14.3	3.1%
TREATMENT2_REP2	95.3%	13.9	51.1%	46%	15.1	3.6%

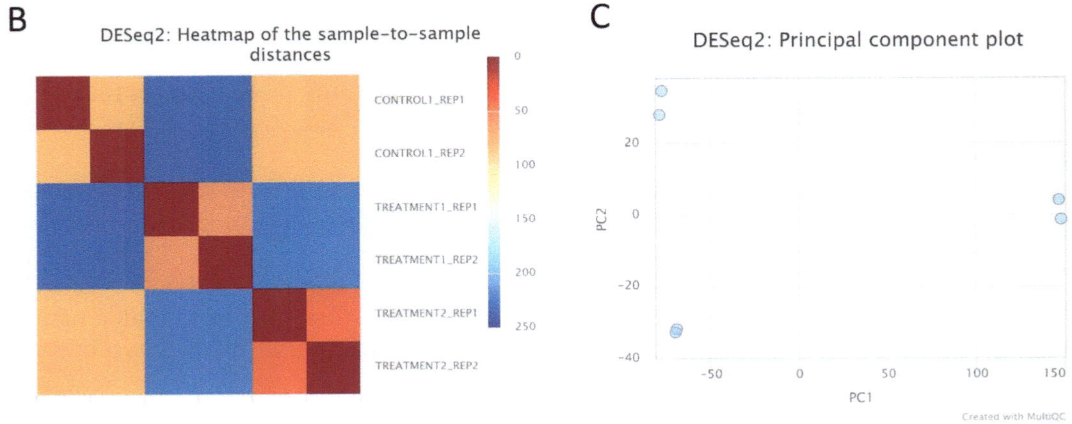

Fig. 2 Examples of the output generated by MULTIQC. (**a**) Complete statistics report of the RNA-seq samples, including number of raw reads before and after trimming adapters and the alignment rates of the samples. (**b**) Sample similarity computed as the Euclidean distances between DESeq2 rlog values for each sample. (**c**) PCA plot between samples in the study

3. A matrix of RNA-seq quantification, where columns correspond to samples and rows to genes, is generated in "`wgcna-modules_testdata/rna_seq/salmon/salmon.merged.gene_tpm.tsv`." This matrix is used as input in the WGCNA clustering procedure in the next step.

3.2 WGCNA Clustering Module

WGCNA performs a correlation network analysis of large, high-dimensional RNA-seq datasets. This unsupervised clustering approach groups similarly expressed genes into gene modules of highly correlated genes that can be summarized in a module eigengene (*see* **Note 11**) and which can be further interpreted for functional analysis, including search of enriched TFBS [13]. Therefore, this method is suitable for managing and summarizing large datasets in a limited number of gene modules/groups with biological meaning (*see* **Note 12**).

3.2.1 Input Data Files

We offer the possibility of running WGCNA with different input alternatives to provide the user with a higher versatility to analyze their own data. The straightforward analysis involves the TPM (tags per million)-normalized counts matrix "`salmon.merged.gene_tpm.tsv`" generated in Subheading 3.1, **step 2** ("TPM mode"). Alternatively, counts can be transformed to Log_2Ratio between experimental and control samples when provided ("Ratio mode"). In addition, the analysis can be performed with the complete gene set (default mode) or with a prefiltered list of genes defined by the user ("Selected genes mode"). Input files required in each case are as follows:

1. "TMP" and "Ratio" modes
   - "`/path/to/salmon_folder`." The path to the folder with Salmon output, including raw quantified reads and TPM- normalized expression matrix generated in Subheading 3.1, **step 2**.
   - "`samplesheet_wgcna.csv`." This file defines the samples and their replicates for WGCNA analysis. It is used to compute the mean of the replicates and to match samples with RNA-seq output (*see* **Note 13**).
   - (Optional "Selected genes mode") "`genes.txt`." This file lists the genes to restrict clustering analysis. If not provided, results in WGCNA are performed on the complete gene set as default (*see* **Note 14**).
2. "Ratio" mode.
   - "`contrasts_wgcna.csv`." This file defines contrasts between experimental and control samples for calculating the corresponding Log_2Ratio values to be used as input in WGCNA analysis (*see* **Note 15**).

3.2.2 Running WGCNA

WGCNA is available as an R package, and here we take advantage of a wrapped WGCNA pipeline in Nextflow to simplify the complete workflow. We refer to the original R package for a fully controlled step-by-step procedure and fine-tuning of the parameters [13]. Complementary to WGCNA, and integrated in the pipeline, we offer the possibility of performing a differential expression analysis using DESeq2 [14] for further automatic selection of differentially expressed genes (DEGs) between experimental and control samples.

1. Install the pipeline from the GitHub repository by running.

```
git clone https://github.com/roldanjg/nf-core-wgcnamodules.
git
```

2. Run the command below to perform WGCNA. Do not forget to replace folder and file names by your own data names (*see* **Note 16** and **Note 17**).

```
nextflow run nf-core/wgcnamodules --input wgcnamodules_test-
data/samplesheet_wgcna.csv --contrast wgcnamodules_testdata/
contrasts_wgcna.csv --norm [TPM/ratio] --salmon_dir wgcnamo-
dules_testdata/rna_seq/salmon --max_cpus 6 --max_memory 16.GB
--diff_exp_genes false
```

3. Add the parameter "`--genes genes.txt`" to run the "Selected genes mode" and to obtain the co-expression modules from a custom list of genes of interest, instead of the complete gene set as default:

```
nextflow run nf-core/wgcnamodules --input wgcnamodules_test-
data/samplesheet_wgcna.csv --contrast wgcnamodules_testdata/
contrasts_wgcna.csv --norm [TPM/ratio] --salmon_dir wgcnamo-
dules_testdata/rna_seq/salmon --max_cpus 6 --max_memory 16.GB
--genes genes.txt --diff_exp_genes false
```

4. Add the argument "`--diff_exp_genes true`" to the command to perform DESeq2 analysis to filter for DEGs (*see* **Note 18**). Criteria for gene filtering are defined with the arguments "`--FDR`" and "`--logratio`," set to 0.05 and 1, respectively, as defaults. Note that the argument "`--genes genes.txt`" is removed in this mode. Run the command as follows:

```
nextflow run nf-core/wgcnamodules --input wgcnamodules_test-
data/samplesheet_wgcna.csv --contrast wgcnamodules_testdata/
contrasts_wgcna.csv --norm [TPM/ratio] --salmon_dir wgcnamo-
dules_testdata/rna_seq/salmon --max_cpus 6 --max_memory 16.GB
--diff_exp_genes true --FDR 0.05 --logratio 1
```

3.2.3 Output Data Files

1. The lists of genes included in each module are generated in the folder "`wgcnamodules_testdata/wgcnamodules/wgcna_output_files/tdthub_modules`." These lists are to be used in the next step for TFBS prediction, but they can also be analyzed in any other functional analysis (*see* **Note 19**).
2. "`module_eigengene_values.csv`." A matrix containing the eigengene values for all the modules is stored in the folder "`wgcnamodules_testdata/wgcnamodules/wgcna_output_files`" (*see* **Note 11**).
3. Some plots illustrating the clustering results and the relationships among the modules in PDF format under the folder "`wgcnamodules_testdata/wgcnamodules/wgcna_output_files`".
   - "modules_dendogram_before_after." Plot with cluster dendrogram and module colors before and after merging close modules, as measured by the correlation of their eigengenes, set at 0.25 as default parameter (Fig. 3).

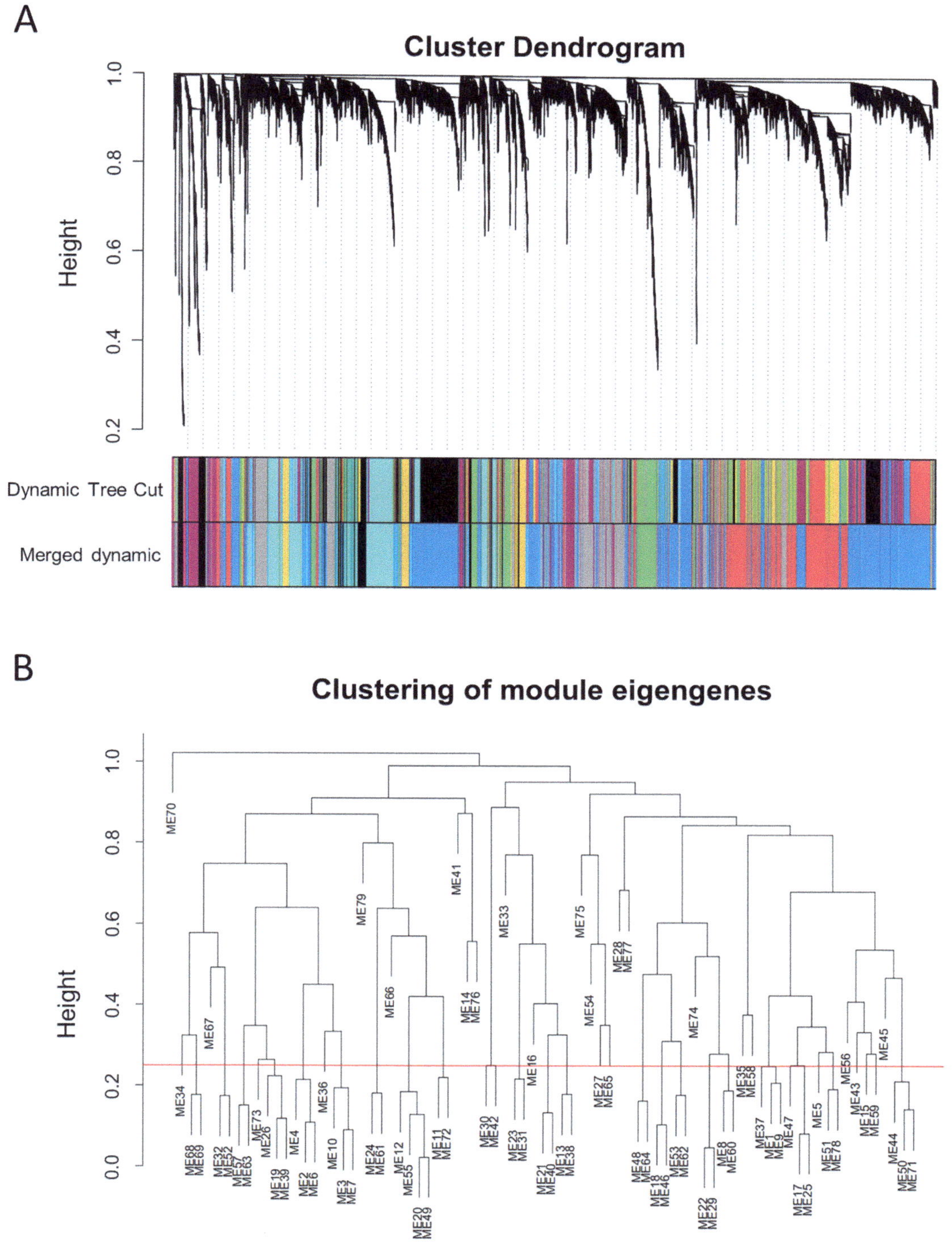

Fig. 3 Examples of the output generated by WGCNA. (**a**) Cluster dendrogram of the gene modules generated before ("Dynamic Tree Cut") and after ("Merged dynamic") merging close modules. (**b**) Clustering of module eigengenes. Horizontal red line indicates the threshold for merging closed modules, measured as the distance *d* = *1- correlation*, set at 0.25 as default

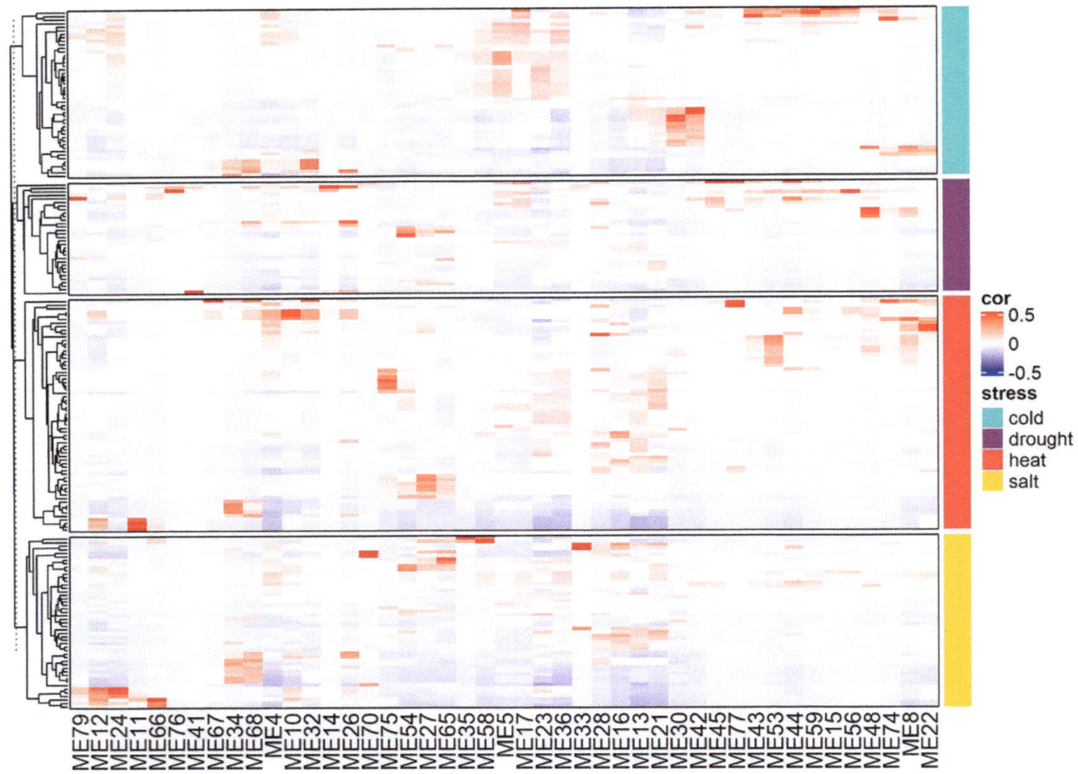

Fig. 4 Heatmap of eigengene values generated after WGCNA clustering. Columns represent gene clusters (or modules) and rows correspond to RNA-seq experiments, which have been grouped by abiotic stress type in this example

- "modules_distance." Clustering of module eigengenes (Fig. 3)
- "module_trait_relationship." A basic heatmap showing modules-traits (i.e., samples) relationship, generated from the eigengene matrix (Fig. 4).

4. (Optional "`--diff_exp_genes true`" mode) DE report. Interactive HTML summary of the differential expression analysis performed by DESeq2. The report is saved in the folder "`wgcnamodules_testdata/wgcnamodules/diff_output_files/report`" (Fig. 5).
5. (Optional "`--diff_exp_genes true`" mode) "`diff_selected_genes.txt`." Gene list filtered by FDR and Log_2Ratio thresholds obtained from the differential expression analysis.

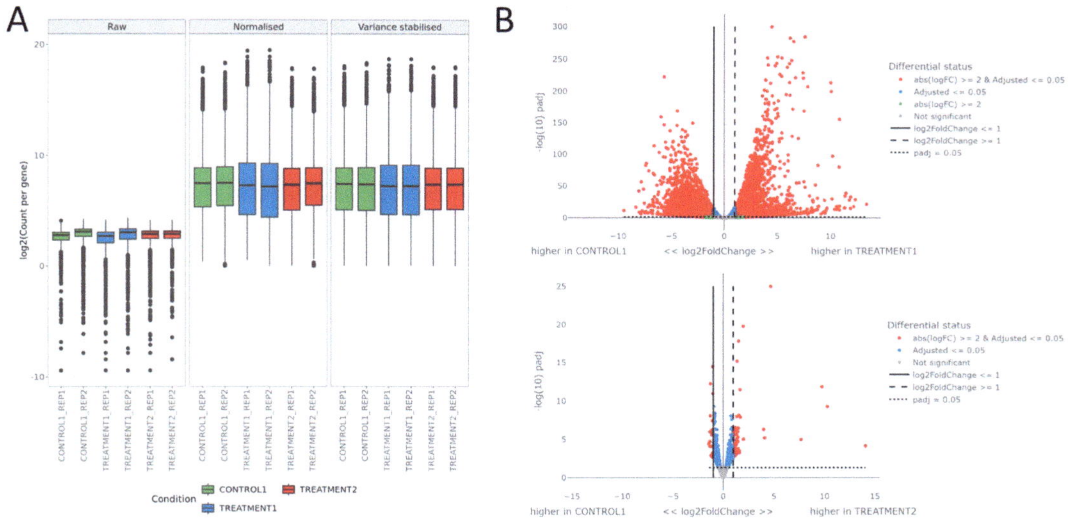

Fig. 5 Examples of the output generated by DE module with DESeq2. (**a**) Read distribution before and after normalization. (**b**) Volcano plots highlighting differentially expressed genes between treated (two different conditions) and control plants

3.3 Analysis of TFBS Enrichment

This step aims to identify enriched TFBS in the modules generated in Subheading 3.2. There are several tools available to predict enriched TFBS from lists of genes of interest, and here we are going to focus on TDTHub [9]. TDTHub is a web-based tool with quick and intuitive access to study the transcriptional control of gene sets that belong to the same metabolic or developmental pathway, or genes exhibiting co-expression or co-regulation in network analysis, which is of interest to this pipeline. TDTHub includes TFBS information for 60 plant species, covering the most popular species and crops, which makes it a very useful tool for studying transcriptional networks in plants.

3.3.1 Input Data and Access to TDTHub

1. Gene lists to be used as input are generated in Subheading 3.2.3 and stored in the "`tdthub_modules`" folder.
2. Navigate the web-based tool's homepage at http://acrab.cnb.csic.es/TDTHub/ and select your species of interest (*see* **Note 20**).
3. Paste the list of genes to analyze in the pop-up window. The "Example format" button will show the Gene ID format required. Then, the website will guide you through a drop-down menu to facilitate the full analysis.
4. Select the TFBS mapping method: FIMO or CB (*see* **Note 21**).
5. Select the size and type of the regulatory regions to which the analysis will be restricted: 1 or 3 kb upstream of the TSS for FIMO and 1, 3, or 5 kb in CB. The latter also permits adding

introns and downstream regions of the genes to the analysis (*see* **Note 22**).

6. Select the S-score threshold. The option "No threshold" will display all the TFBSs (*see* **Note 23**).

3.3.2 Results and Output Data

1. The results table will display the TFBSs along with their enrichment stats in the input gene list. Download the results as ".tsv" files (*see* **Note 24**).
2. After repeating this analysis for all the modules of interest, the resulting ".tsv" files can be uploaded to an interactive heatmap plot accessible from the top bar menu. This heatmap provides a visual representation of the variation in TFBS across different gene modules (Fig. 6).

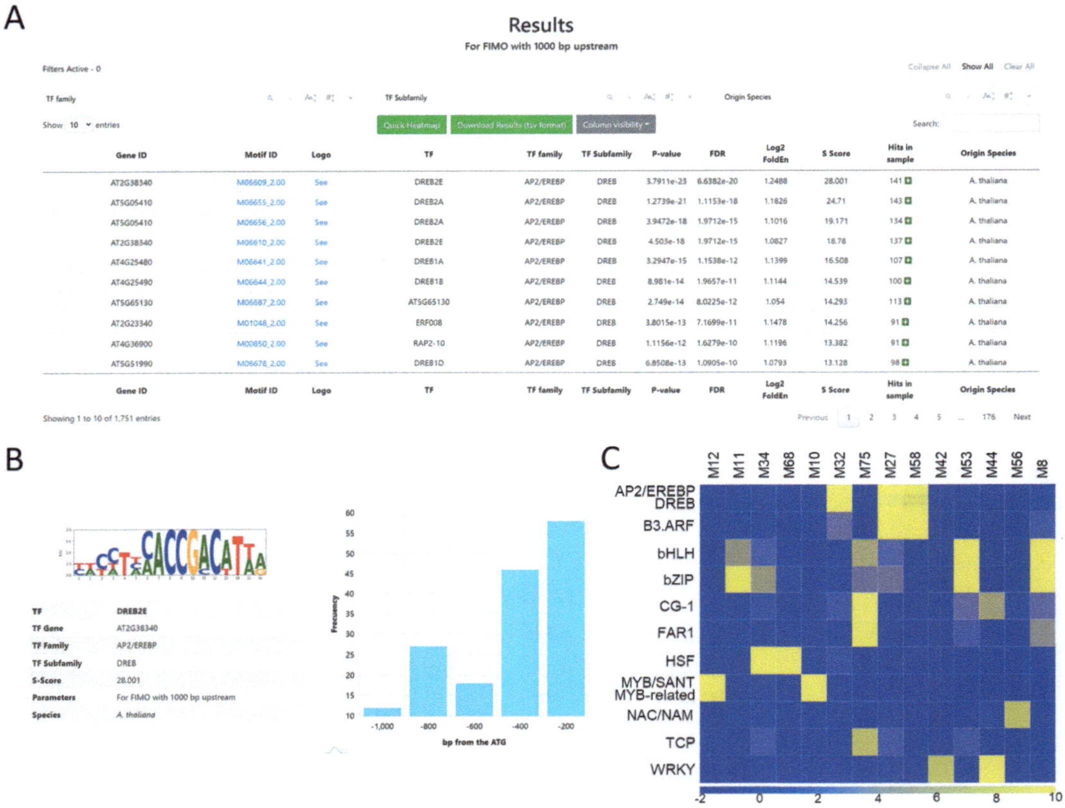

Fig. 6 Examples of the output generated by TDTHub. (**a**) Screenshot of the results table with the top 10 TFBS obtained for module M32 and analyzed with FIMO in 1 kb-sized promoters. TFBS have been sorted based on their S-score. (**b**) Detail of the highest significant TFBS and distribution of motifs along the promoters of the genes in the module. (**c**) Heatmap showing enrichment (S-score) in some gene clusters identified in Fig. 4 selected for this example. In this plot, only top 10-scoring motifs from each cluster were selected

4 Notes

1. For the sake of simplicity and reproducibility, this pipeline has as requirement a personal computer without any specific software already installed. Computational requirements may depend on the amount of data to be analyzed and the time limitations of the project. In the example shown in this guide, we used a laptop with Intel(R) Core (TM) i7-10510U CPU at 2.30 GHz, 16 Gb RAM, and 100 GB of free SSD space with Windows 10 distribution for the method description. Large-scale datasets have been analyzed on a server Intel(R) Xeon (R) Gold 6240R CPU at 2.40GHz (96 threads), 128 Gb RAM, and 7 TB of SSD with Ubuntu 20.04.6.
2. Windows users will need Windows Subsystem Linux (WSL). If not preinstalled, follow these steps:
   - Open PowerShell or Windows Command Prompt in administrator mode by right-clicking and selecting "Run as administrator."
   - Type "wsl --install --distribution Ubuntu 22.04 LTS," execute, and restart the computer. This will install the Ubuntu distribution indicated.
   - To edit, move or modify your files inside Ubuntu—you can use the Windows 10/11 file explorer.
3. Open the Ubuntu terminal to access the command line console for the base folder: `user@computername:~$`. Type "`sudo apt update && sudo apt upgrade`."
4. Miniconda 3 is a free minimal installer for conda that only includes Python, conda, their dependencies, and a small number of other useful packages. Install Miniconda for Linux by typing:

```
mkdir -p ~/miniconda3
wget https://repo.anaconda.com/miniconda/Miniconda3-latest-Linux-x86_64.sh -O ~/miniconda3/miniconda.sh
bash ~/miniconda3/miniconda.sh -b -u -p ~/miniconda3
rm -rf ~/miniconda3/miniconda.sh
conda config --set auto_activate_base false
```

   If you are using Mac/Windows, or if the "`wget`" command is not working, change to the appropriate installer from conda's official project site (https://docs.anaconda.com/free/miniconda/). It is recommended to install conda's latest version.

   Change the source packages channels priority by typing.

```
conda config --add channels defaults.
conda config --add channels bioconda.
conda config --add channels conda-forge.
conda config --set channel_priority strict.
```

5. Nextflow is a scientific workflow system predominantly used for bioinformatic data analyses that enables scalable and reproducible scientific pipelines using software containers or virtual environments. It allows the integration of code written in the most common scripting languages. More information about Nexflow can be found at https://www.nextflow.io/. Nf-core is a community effort to collect a curated set of analysis pipelines built using Nextflow (https://nf-co.re/) [15].
   - Install Nextflow and nf-core inside a conda virtual environment:

```
conda create --name nf-core python = 3.11 nf-core = 2.13.1
nextflow = 23.10.1
```

   - Activate the environment to perform any of the analysis in this method:

```
conda activate nf-core
```

   - To test the pipeline, run the following:

```
nextflow run nf-core/rnaseq -r 3.14.0 -profile test,conda
--outdir test_rnaseq.
```

   A message "`Pipeline completed successfully`" should appear.

6. Reference sequence and annotation files can be downloaded from several sources. Ensemble Plants (https://plants.ensembl.org/index.html) contains reference files for 151 plant species, whereas Phytozome v13 (https://phytozome-next.jgi.doe.gov/) hosts 395 assembled and annotated genomes. It is worth noting that files from different sources do not necessarily correspond to identical genome versions and it is recommended to download reference files from the same source.
7. RNA-seq data can correspond to own experimental results in FASTQ format generated by a sequencing provider or can be directly downloaded from publicly available databases. There are several repositories of high throughput sequencing data that store and curate nucleotide sequences under the FAIR principles to promote data reuse and scientific progress [16]. The most popular databases include SRA (https://www.ncbi.nlm.nih.gov/sra), ENA (https://www.ebi.ac.uk/ena/),

NGDC (https://ngdc.cncb.ac.cn), and DDBJ (https://www.ddbj.nig.ac.jp/) [17–20]. RNA-seq data can be either single end (SE), where just a single FASTQ file per sample is generated, or paired end (PE), which involves two files with identical names but differing in a suffix, usually "_1" and "_2," to differentiate both ends.

8. Examples and templates for the files needed as inputs in all the steps can be downloaded with the following command:

```
git clone https://github.com/roldanjg/wgcnamodules_testdata.
git
```

Then, you can test the pipeline by running this command:

```
nextflow run nf-core-wgcnamodules -profile test_full,conda --
outdir test_wgcna
```

9. The metadata samplesheet must fit to the format accepted by the pipeline, according to the usage documentation (https://nf-co.re/rnaseq/3.14.0/docs/usage). It is structured as a comma-separated file comprising four columns, where the rows correspond to the samples to be analyzed, and needs to be located in the same folder as the FASTQ files, with the file extension ".csv." An example of the samplesheet named "samplesheet_rnaseq.csv" is provided below:

```
sample,fastq_1,fastq_2,strandedness
CONTROL1_REP1,CNTRL1_REP1_1.fastq.gz,CNTRL1_REP1_2.fastq.gz,
auto
CONTROL1_REP2,CNTRL1_REP2_1.fastq.gz,CNTRL1_REP2_2.fastq.gz,
auto
```

The last column is set to "auto" as the pipeline can automatically detect library strandedness. If RNA-seq data are SE, leave fastq_2 column empty:

```
sample,fastq_1,fastq_2,strandedness
CONTROL1_REP1,CNTRL1_REP1_1.fastq.gz,,auto
CONTROL1_REP2,CNTRL1_REP2_1.fastq.gz,,auto
```

10. This configuration is tailored to accommodate for the minimum computer requirements. Depending on the amount of data to analyze and the computing capacity, the entire pipeline may require several hours to run. Parameters "`--max_cpus 6 --max_memory 8.GB`" can be adjusted if additional resources are available for faster analysis.

11. The module eigengene is defined as the first principal component of a given module. It can be considered a representative of the gene expression profiles in a module [13].
12. A minimum of three different experimental conditions is required for WGCNA analysis [13].
13. We recommend using the format below as a template for the "`samplesheet_wgcna.csv`" file:

```
sample,condition,replicate
CONTROL1_REP1,CONTROL1,1
CONTROL1_REP2,CONTROL1,2
TREATMENT1_REP1,TREATMENT1,1
TREATMENT1_REP2,TREATMENT1,2
TREATMENT2_REP1,TREATMENT2,1
TREATMENT2_REP2,TREATMENT2,2
```

where the columns correspond to

- "sample": Same name as the "samplesheet_rnaseq.csv" sample column
- "condition": Name of the treatment, genotype, or group that defines an experimental condition with one or multiple replicates
- "replicate": number of the biological replicate

14. The format of "`genes.csv`" file must be one gene per line.
15. Use the format below as a template for "`contrasts_wgcna.csv`":

```
contrast,variable,control,target
TREATMENT1_vs_CONTROL1,condition,CONTROL1,TREATMENT1
TREATMENT2_vs_CONTROL1,condition,CONTROL1,TREATMENT2
```

where the columns correspond to

- contrast: A custom name used to identify the contrast
- variable: The name of the column from "samplesheet_wgcna.csv" file that contains the condition IDs
- control: The base/reference level for the contrast,
- target: The target/non-reference level for the comparison.

16. Select between the options "--norm TPM" or "--norm ratio" to carry out the WGCNA analysis with the "TMP" or "Ratio" modes, respectively.
17. This pipeline has been tested with *Arabidopsis* samples, and the parameters have been optimized manually to extract the maximum amount of information for the number of genes and the

characteristics of a large number of samples, based on the default modeling properties established in [13]. We do not provide any further tuning of WGCNA parameters in this pipeline, but any modifications could be done manually by editing the R code inside the "bin" folder of the pipeline.

18. Automated analysis of differential expression is optional, but we have integrated this step as an independent module to enhance the capabilities of this pipeline.
19. The lists of gene modules are generated in a multistep process. In the first step, an adjusted matrix of Pearson correlation coefficients between gene pairs is used to obtain the topological overlap matrix (TOM). The second step involves hierarchical clustering on the TOM values to identify gene modules. Then, Module Membership (MM) for each gene is calculated by correlating its gene expression profile with the module's eigengene of a given module [13]. Finally, genes are reassigned to the module with the highest MM value.
20. A complete tutorial of TDTHub is available at http://acrab.cnb.csic.es/TDTHub/tutorial.
21. Find Individual Motif Occurrences (FIMO) and Cluster-Buster (CB) are two different mapping algorithms used in the precomputation of TFBS. FIMO assigns a false discovery rate (FDR)-corrected P-value to motif occurrences [21], whereas CB finds dense clusters of motifs in input sequences [22]. Selection of one or another method may generate slightly different results, and we recommend selecting FIMO as a starting point.
22. Some reports have demonstrated that 5 kb upstream regions, including introns and downstream regions of the genes, may improve identification of TF-target genes with CB [23]. In our experience, these do not have a significant effect when using FIMO as the mapping algorithm.
23. Significance Score (S-score) is defined as the product $-\,log10\,(P) * FE$, and represents the significance of TFBS enrichment. False Positive thresholds for S-scores at 5% or 1% have been calculated empirically for each species, 1% being the most restrictive.
24. The results table from TDTHub is interactive and allows sorting text searching and custom rendering for different categories. The nucleotide coordinates of a given TFBS in target genes can be obtained, as well as a basic bar graph showing the distribution in the promoter (Fig. 6).

Acknowledgments

This work has been funded by grants from the Spanish Ministry of Economy (MINECO) and fondos FEDER (PID2020-119451GB-I00 and PID2023-149400NB-I00). J.M.F.Z. acknowledges financial support from the Spanish State Research Agency, AEI/10.13039/501100011033, through the "Severo Ochoa" Programme for Centres of Excellence in R&D (SEV-2017-0712 and CEX2023-001386-S).

Conflict of Interest The authors have no conflicts of interest to declare that are relevant to the content of this chapter.

References

1. Franco-Zorrilla JM, Solano R (2017) Identification of plant transcription factor target sequences. Biochim Biophys Acta - Gene Regul Mech 1860:21–30. https://doi.org/10.1016/j.bbagrm.2016.05.001
2. Sullivan AM, Arsovski AA, Lempe J et al (2014) Mapping and dynamics of regulatory DNA and transcription factor networks in A. thaliana. Cell Rep 8:2015–2030. https://doi.org/10.1016/j.celrep.2014.08.019
3. Pajoro A, Madrigal P, Muino JM et al (2014) Dynamics of chromatin accessibility and gene regulation by MADS-domain transcription factors in flower development. Genome Biol 15: R41. https://doi.org/10.1186/gb-2014-15-3-r41
4. Buenrostro JD, Giresi PG, Zaba LC et al (2013) Transposition of native chromatin for fast and sensitive epigenomic profiling of open chromatin, DNA-binding proteins and nucleosome position. Nat Methods 10:1213–1218. https://doi.org/10.1038/nmeth.2688
5. Weirauch MT, Yang A, Albu M et al (2014) Determination and inference of eukaryotic transcription factor sequence specificity. Cell 158:1431–1443. https://doi.org/10.1016/j.cell.2014.08.009
6. Franco-Zorrilla JM, López-Vidriero I, Carrasco JL et al (2014) DNA-binding specificities of plant transcription factors and their potential to define target genes. Proc Natl Acad Sci 111: 2367–2372. https://doi.org/10.1073/pnas.1316278111
7. O'Malley RC, Huang SC, Song L et al (2016) Cistrome and Epicistrome features shape the regulatory DNA landscape. Cell 165:1280–1292. https://doi.org/10.1016/j.cell.2016.08.063
8. Rauluseviciute I, Riudavets-Puig R, Blanc-Mathieu R et al (2023) JASPAR 2024: 20th anniversary of the open-access database of transcription factor binding profiles. Nucleic Acids Res 52:D174. https://doi.org/10.1093/nar/gkad1059
9. Grau J, Franco-Zorrilla JM (2022) TDTHub, a web server tool for the analysis of transcription factor binding sites in plants. Plant J 111: 1203–1215. https://doi.org/10.1111/tpj.15873
10. Krueger F, James F, Ewels P, et al (2021) FelixKrueger/TrimGalore: v0.6.7 - DOI via Zenodo
11. Patro R, Duggal G, Love MI et al (2017) (2017) Salmon provides fast and bias-aware quantification of transcript expression. Nat Methods 144(14):417–419. https://doi.org/10.1038/nmeth.4197
12. Soneson C, Love MI, Robinson MD (2016) Differential analyses for RNA-seq: transcript-level estimates improve gene-level inferences. F1000Research 4:1521. 10.12688/f1000research.7563.2
13. Langfelder P, Horvath S (2008) WGCNA: an R package for weighted correlation network analysis. BMC Bioinf 9:559. https://doi.org/10.1186/1471-2105-9-559
14. Love MI, Huber W, Anders S (2014) Moderated estimation of fold change and dispersion for RNA-seq data with DESeq2. Genome Biol 15:550. https://doi.org/10.1186/s13059-014-0550-8
15. Ewels PA, Peltzer A, Fillinger S et al (2020) The nf-core framework for community-curated bioinformatics pipelines. Nat Biotechnol 383(38):276–278. https://doi.org/10.1038/s41587-020-0439-x

16. Wilkinson MD, Dumontier M, Aalbersberg IJJ et al (2016) The FAIR Guiding Principles for scientific data management and stewardship. Sci Data 3:160018. https://doi.org/10.1038/sdata.2016.18
17. Katz K, Shutov O, Lapoint R et al (2022) The sequence read archive: a decade more of explosive growth. Nucleic Acids Res 50:D387–D390. https://doi.org/10.1093/NAR/GKAB1053
18. Burgin J, Ahamed A, Cummins C et al (2023) The European nucleotide archive in 2022. Nucleic Acids Res 51:D121–D125. https://doi.org/10.1093/NAR/GKAC1051
19. Fukuda A, Kodama Y, Mashima J et al (2021) DDBJ update: streamlining submission and access of human data. Nucleic Acids Res 49: D71–D75. https://doi.org/10.1093/NAR/GKAA982
20. Bao Y, Zhang Z, Zhao W et al (2024) Database resources of the National Genomics Data Center, China National Center for bioinformation in 2024. Nucleic Acids Res 52:D18–D32. https://doi.org/10.1093/NAR/GKAD1078
21. Grant CE, Bailey TL, Noble WS (2011) FIMO: scanning for occurrences of a given motif. Bioinformatics 27:1017–1018. https://doi.org/10.1093/bioinformatics/btr064
22. Frith MC, Li MC, Weng Z (2003) Cluster-buster: finding dense clusters of motifs in DNA sequences. Nucleic Acids Res 31:3666–3668. https://doi.org/10.1093/nar/gkg540
23. Kulkarni SR, Vaneechoutte D, Van de Velde J, Vandepoele K (2018) TF2Network: predicting transcription factor regulators and gene regulatory networks in Arabidopsis using publicly available binding site information. Nucleic Acids Res 46:e31. https://doi.org/10.1093/nar/gkx1279

Chapter 14

Web-Based Discovery of Regulatory Motifs in Non-model Plants

Najla Ksouri, Ammar Abdalrahem, Francesc Montardit-Tarda, Jaime A. Castro-Mondragón, Jacques van Helden, Yolanda Gogorcena, and Bruno Contreras-Moreira

Abstract

This protocol describes step-by-step how to discover and annotate DNA motifs in regulatory regions of clusters/modules of co-expressed genes in plants, including non-model species. It also explains how to control the significance of the results and how to associate the discovered motifs with potentially binding transcription factors (TFs). These tasks can all be done in a web browser. While it uses the plant-dedicated mirror of the Regulatory Sequence Analysis Tools (RSAT, http://plants.rsat.eu), it can be used on any organism hosted at any RSAT mirror.

Key words Plant genome, Non-model species, DNA motif discovery, Proximal promoter, Cluster, Module

1 Introduction

Transcriptome data have been extensively used as a proxy for genetic regulation in many organisms, as the analysis of genome-wide transcriptional profiles under different treatments uncovers clusters of genes with correlated behaviors, which may result from direct or indirect co-regulation. A classical application of this approach was done by Beer and coworkers with yeast microarray datasets obtained in a variety of experimental conditions [1]. In that experiment, expression data-mining was demonstrated to be an effective strategy for finding groups of genes that share regulatory mechanisms and functional annotations.

Previous studies have unveiled that the outcome of these approaches largely depends on the genomic background of the species under study. For instance, Sand and collaborators [2] reported that the significance of DNA motifs discovered in baker's yeast gene promoters is much higher for regulons than for random

José M. Franco-Zorrilla and Carmen Martín-Pizarro (eds.), *Plant Transcription Factors: Methods and Protocols*, Methods in Molecular Biology, vol. 2985, https://doi.org/10.1007/978-1-0716-4972-5_14,

gene sets of the same sizes. However, for human promoters, the signal-to-noise ratio is almost null, because random gene sets give highly significant motifs due to heterogeneities in promoter compositions and biases due to repetitive elements. For metazoans, it is thus a real challenge to distinguish bona fide motifs from noise [2]. These observations suggest that motif discovery on sequence clusters faces intrinsic properties of the genomes under study, regardless of the software used.

Among plants, these strategies have so far been tested on the model *Arabidopsis thaliana*, and they have been successfully applied to the identification of novel *cis*-regulatory elements validated with synthetic promoters [3]. Yet, with the exception of this model, these experiments have not been possible in other plants until recently. In spite of this, the growing list of available plant genomes encourages these analyses in combination with expression profiles, as in the recent work of Yu and collaborators in maize [4], provided that these factors are considered:

- Plant genomes are rich in repetitive elements (RE) distributed along the genome [5], which pose particular problems for motif discovery statistics (violation of the independence assumption). Moreover, recent work on *Brachypodium distachyon*, a model species for grasses, suggests that accessory genes, found only in some populations, accumulate more RE in their upstream regions than core genes (Fig. 1).
- Current genome assemblies at RSAT::Plants range from 119.7 Mb (*A. thaliana*) to 14.54Gb (bread wheat, *Triticum aestivum*). The quality of these annotated assemblies and their RE content is variable [7].
- Although generally the margins of promoter regions are usually not annotated, recent work suggests that the interval spanning from −500 bp to +200 bp, relative to the transcription start site (TSS), works well in peach (*Prunus persica*) and other plants [8].

This tutorial presents a comprehensive step-by-step protocol for the identification and annotation of DNA motifs within clusters of upstream sequences for species supported on the RSAT::Plants server (Table 1). It consolidates and builds upon prior protocols [8, 9] to offer an updated and enhanced methodology. RSAT::Plants also integrates footprintDB, a curated collection of regulatory DNA motifs and their cognate binding TFs [10]. Plant species in footprintDB with two or more TFs are listed in Table 2.

2 Materials

This protocol requires

- A computer with any web browser installed.
- Sets of gene clusters/modules derived from any of the species currently supported at RSAT::Plants (http://plants.rsat.eu; *see*

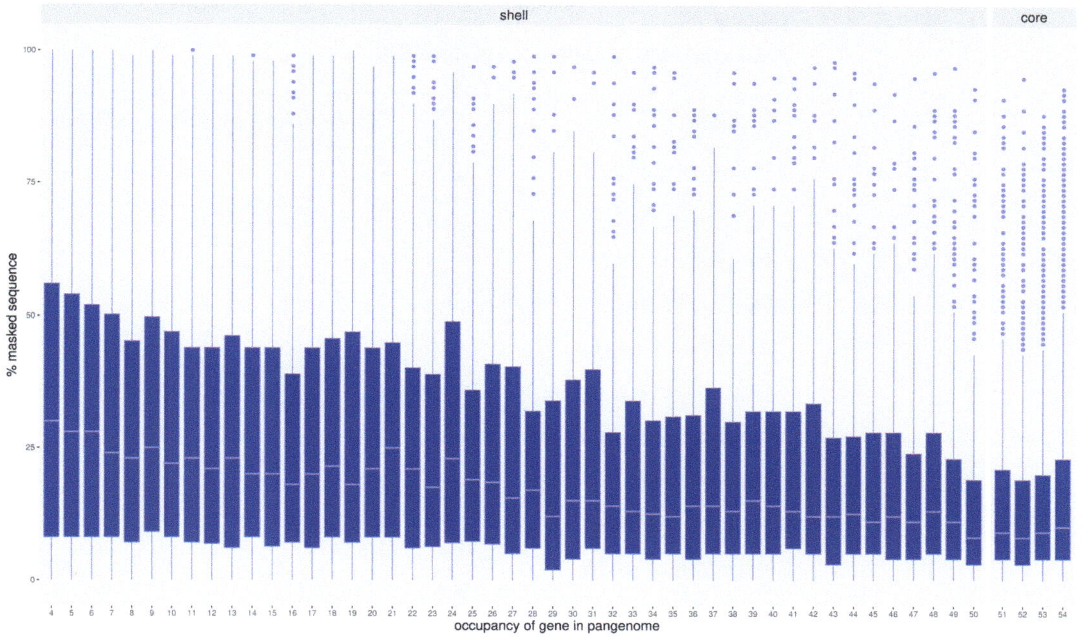

Fig. 1 Proportion of repeated sequences masked in upstream regions of genes from *Brachypodium distachyon* as a function of the number of populations in which each gene is found (occupancy). The total number of populations sampled in this experiment was 54, adapted from [6]

Table 1
Source of annotated plant genomes in RSAT::Plants. As of January 8, 2024, there were a total of 108 supported datasets

Source	Total datasets
Ensembl	88
NCBI	7
DOE-joint genome institute	7
Others	6

Note that "others" include Genome Database for Rosaceae (GDR), Senckenberg Biodiversity and Climate Research Centre (SBiKF), Tokyo Institute of Technology (TITECH), Sol Genomics Network (SGN), and the International Potato Genome Sequencing Consortium (PGSC). Check up-to-date data at https://rsat.eead.csic.es/plants/supported-organisms.cgi

Note 1). In previous work, we have produced sets of clusters based on gene expression data. In our experience, simple hierarchical clustering demonstrated reasonable effectiveness, but it requires extra criteria to help define the final number of clusters. Weighted Correlation Network Analysis (WGCNA) provides superior performance and is supported by a well-documented methodology. You can read about the performance of both approaches at [8].

Table 2
Plant contents in database footprintDB

Species	Transcription factors	DNA motifs
Hordeum vulgare	2	4
Lotus japonicus	2	2
Manihot esculenta	2	2
Populus deltoides x P. euramericana	2	1
Antirrhinum majus	3	10
Medicago truncatula	3	3
Brachypodium distachyon	4	4
Pisum sativum	4	3
Triticum aestivum	4	5
Nicotiana tabacum	5	6
Vitis vinifera	6	6
Malus x domestica	7	10
Glycine max	8	25
Solanum lycopersicum	9	14
Arabidopsis lyrata	11	21
Cannabis sativa	19	19
Physcomitrium patens	25	31
Ostreococcus tauri	27	26
Oryza sativa	42	63
Zea mays	59	120
Arabidopsis thaliana	945	2454

As of January 8, 2024, there were a total of 1242 transcription factors and 2878 DNA motifs. The table only shows species with two or more transcription factors. The complete data files can be downloaded from https://footprintdb.eead.csic.es/download. The contents include original and manually curated data from several sources (*see* https://footprintdb.eead.csic.es/index.php?databases for their contributions and citations). Separate TRANSFAC-format motif collections and licenses are also available at https://github.com/rsa-tools/motif_databases

The first lines of module M9 from [8] are as follows, with each line corresponding to the identifier of a gene model in the peach genome, used for that study:

PRUPE_8G027100

PRUPE_8G036800

PRUPE_8G052100

PRUPE_8G060700
PRUPE_8G075000
PRUPE_8G082700
PRUPE_8G134300
PRUPE_8G137300
PRUPE_8G150300
PRUPE_8G152100
...

3 Methods

The following protocol enumerates the steps required to discover DNA motifs, based on the overrepresentation of k-mers (oligonucleotides of length k) and dyads (spaced pairs of oligonucleotides), in modules/clusters of upstream sequences (*see* **Note 2**). The protocol comprises two stages, starting with the analysis of co-expressed genes followed by the examination of random clusters as a negative control (*see* **Note 3**). This two-stage approach ensures that the identified motifs are not simply artifacts of the analysis. By completing both stages, a nuanced understanding of the results can be achieved, supporting informed decisions about their relevance and biological significance. The estimated time to carry out the following steps is approximately 1 h.

3.1 Collecting the Full Set of Promoters for the Genome of Interest

Before analyzing a particular set of gene clusters, we will first retrieve the promoter sequences of all the genes of the organism of interest, which will serve below to estimate the background model.

1. Open a connection to the RSAT::Plants server (http://plants.rsat.eu). On the left-side menu, select "Sequence tools -> retrieve sequence."
2. In the "mandatory inputs" section, choose "Single organism" (*see* **Note 1**).
3. Choose "Genes -> all"; this will retrieve all upstream sequences of the selected genome.
4. Go to "mandatory options" and set the appropriate upstream bounds. Default values are −2000, −1. To replicate the maize study of Yu and collaborators [4], these should be set to "From" −1000 and "To" +200.

   Without prior knowledge of the exact upstream region boundaries, a window-based approach may be adopted. In other terms, the proximal upstream sequences may be defined as windows of variable length, each one examined separately

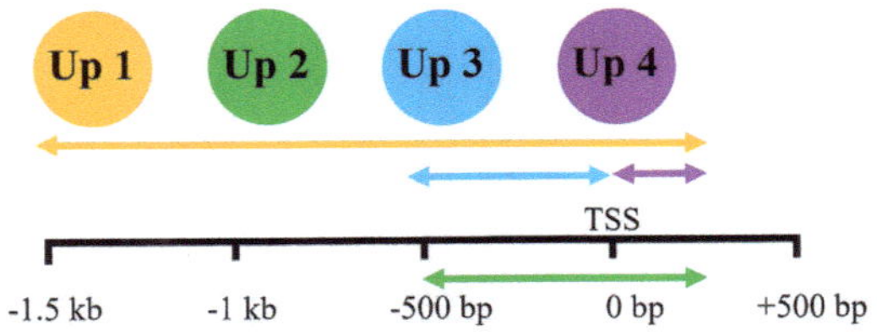

Fig. 2 Delimitation of the upstream sequences/proximal promoters with variable window lengths (Up1, Up2, Up3, and Up4)

(*see* **Note 4**). As described in [8], four promoter intervals can be defined as follows (Fig. 2):

- **Up 1**: ["From" −1500 "To" +200].
- **Up 2**: ["From" −500 "To" +200].
- **Up 3**: ["From" −500 "To" 0].
- **Up 4**: ["From" 0 "To" +200].

For each upstream stretch, a background model of the same size is estimated, and overall, four different background models are generated.

5. We recommend ticking the option "Mask repeats" in the "advanced options," as plant genomes are frequently repeat-rich. However, this option should be skipped if you suspect the TFs of interest bind to repeated sequences.
6. Press "GO" and wait until the result page is displayed (*see* **Note 5**). The last row of the results table should include a URL to the "sequences" file, which must be saved and labeled as "all.fasta.URL."

For the sake of clarity, when analyzing various upstream windows (Up 1 to Up 4), we refer to the background sequences URL as "all-Up1.fasta.URL" to "all-Up4.fasta.URL." These FASTA-format files can also be stored as local files on your computer. However, keep in mind that they might be relatively large, potentially causing uploading issues.

3.2 Analyzing Upstream Sequences of Co-expressed Genes

In this section, we assume that you already have one or more clusters of genes that you want to analyze together.

We will now retrieve the upstream sequences of a cluster of co-expressed genes and use the tool *peak-motifs* to uncover over-represented motifs in their promoters. While this tool was initially conceived to discover motifs in ChIP-seq peaks [11], it was found to work well also with clusters of promoters [9], as done here.

1. Choose a cluster, copy the corresponding gene IDs, and paste them into a new text file that you will store on your computer.

Ensure that newline characters are inserted between genes (*see* **Note 6**).

2. On the left-hand menu of the RSAT server, click on "retrieve sequence" to get a fresh form. In the "mandatory inputs," make sure that the option "Genes -> selection" is activated. Tick "Mask repeats" and set the same size limits as for the whole collection of promoters (*see* **step 4**). Paste the list of IDs of your gene cluster. In case of promoter regions of variable length, redo the same step and make sure to set the corresponding bounds for, respectively, Up 1, Up 2, Up 3, and Up 4 windows.
3. Press "GO" and wait a few seconds until the result page is displayed. Upon inspection of these sequences, you might identify N-masked stretches, which typically correspond to repeats or to unresolved regions in the assembly. Save both "query genes" and "sequences" files locally on your computer. We will refer to them as "cluster.genes" and "cluster.fasta" later on this protocol.

   As described previously, for the upstream boundaries Up 1 to Up 4, we refer to the cluster sequences as "Up1.cluster.fasta" to "Up4.cluster.fasta."
4. Click on the "peak-motifs" button. Please note that in this case, peak-motif will operate under the "differential analysis mode," contrasting cluster sequences (test set) against all upstream sequences from the reference genome (control background). "Peak sequences box" is automatically prefilled with the cluster sequences selected on **steps 1–2**.
5. As Control sequences, paste the "all.fasta.URL" on the "URL of a sequence file available on a Web server" entry form on the right side.
6. Provide a title for this job, such as "cluster1_Up1."
7. Click on "Reduce peak sequences," and in "Cut peak sequences +/−"suppress the default value (500) and leave the field blank. This step prevents the sequences from being clipped to 1Kb (by default, peak-motifs clips the peaks to 500 bp on each side of the peak centers).
8. Click on "Motif discovery parameters." Select two algorithms: "Discover overrepresented words" ("oligo-analysis") and "Discover overrepresented spaced word pairs" ("dyad-analysis"). Check the option "merge lengths for assembly" and uncheck the program "position-analysis" (*see* **Note 7**). Click on "Reporting options," and then set "Origin" to "end" and "Offset" to −200 (*see* **Note 8**). For Up 1 to Up 4, we keep the "Offset" default value set to 0.

9. Click on "Compare discovered motifs with databases," and select the appropriate database which will be used to annotate any discovered motifs. For plant promoters, we recommend "*footprintDB-plants*" ('), but you can also consider other databases such as "*Athamap*," "*ArabidopsisPBM*," and "*JASPAR plants*" (*see* **Note 9**). Alternatively, you can also upload your own collection of DNA motifs in TRANSFAC format. Note that RSAT convert-matrix can be used to change matrix formats as needed.
10. In "output," choose the "email" option and type your delivery address. Given that these are resource-intensive jobs, this ensures that the result page and download link will be sent to your email even if your connection to the server is interrupted. Press "GO" to initiate the analysis.
11. After a few seconds, the server should have uploaded the sequences and displayed a page with the URL of the future result page. You can already click on this link as the result page will be periodically updated to show the progress of the analysis. At the end of the processing, a box will appear at the top of the result page, providing a brief summary of the discovered motifs and links to different sections of the results. Once the job is complete, click on the link "Download all results (peak-motifs_archive.zip)" to save the results on your computer. This archive can be uncompressed later to access the results after their removal from the server (Please note that the results are only available on the server for 72 h after job completion.). We also recommend downloading the full set of discovered motifs, by clicking on the link "Download all matrices (transfac format)" and saving a local file named "cluster.motifs.tf." This file contains all motifs in the form of position-weight matrices (PWMs) in TRANSFAC format.

On the result page, the section entitled "*Discovered motifs (with motif comparison)*" lists the discovered motifs and displays their sequence logos and their distribution along clustered sequences, in addition to top matches with the motif collections selected on **step 9**. As shown on Fig. 3, when the footprintDB collections are used to compare the discovered motifs, you can also easily retrieve information and literature about binding TFs by clicking on their names.

Figure 4 shows an example motif (MA1355.1) curated in footprintDB, with details about the statistical relevance, position-specific scoring matrix, the logo, the original sources, and known binding transcription factors. The "Binding TFs" contents are also clickable. For each TF (T25713 and Q9JFW5 in the example), you can obtain the amino acid sequence, annotated structural domains, predicted protein-DNA interface residues, and related literature.

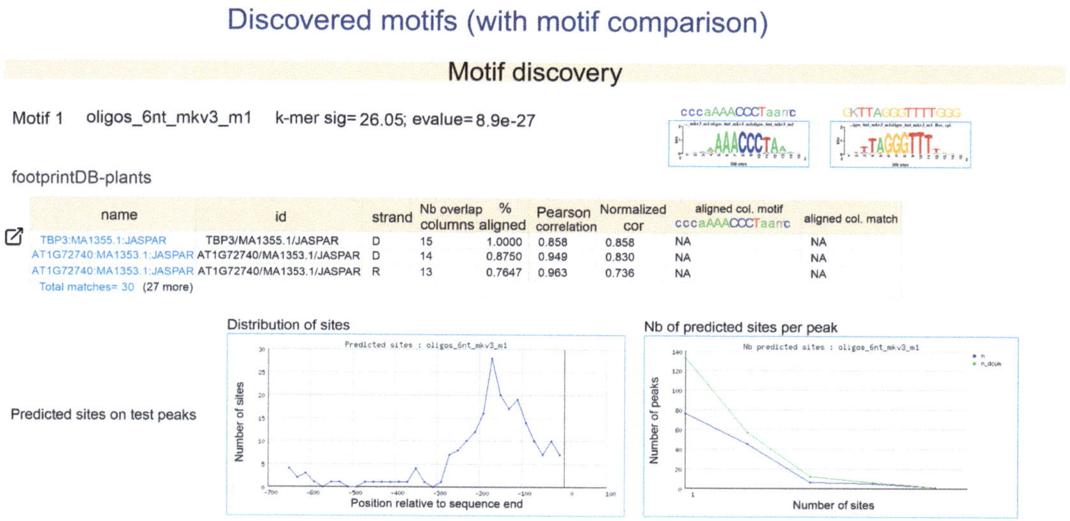

Fig. 3 Comparison of a DNA motif found by peak-motifs and the footprintDB-plants collection. The motif oligos_6nt_mkv3_m1, discovered by algorithm oligo-analysis [12], is compared to two *Arabidopsis thaliana* motifs (TBP3:MA1355.1:JASPAR and AT1G72740:MA1353.1:JASPAR) annotated originally in the JASPAR database [13]. Three different comparisons are summarized sorted in terms of normalized correlation (Ncor), a relative width-normalized Pearson correlation of two PWMs aligned with *matrix-scan* [14]. Note that the column "Name" displays clickable motif names in footprintDB (https:/footprintdb.eead.csic.es). The plots on the bottom show the distribution of motif matches in the input upstream sequences (left) and the number of predicted sites per upstream sequence (right)

3.3 Negative Control: Random Groups of Genes

In this section, we propose a procedure to obtain an empirical estimation of false positive rate, by detecting motifs in the promoters of randomly selected genes.

1. In the left-hand menu, select "Build control sets" -> "random gene selection."
2. Choose the appropriate "Organism."
3. Set the "Number of genes" to match the size of the cluster used above. For convenience, in this tutorial only one random group is generated. However, we emphasize that a robust motif analysis relies on evaluating multiple random clusters. For instance, the examples illustrated in Fig. 5 include 50 negative controls.
4. Press "GO" and, in the "Next step" box, click on "retrieve sequences" at the bottom of the result page. In the new retrieve-seq form, set the other parameters as above: From −1000 to +200, check the "Mask repeats" option. To get reliable results, it is crucial to use the same random group of genes along the upstream windows Up1, Up2, Up3, and Up4 (the same list of genes).

Accessions: **MA1355.1 (JASPAR 2024)**, **M0604 (AthalianaCistrome v4_May2016)**

Names: TBP3, TRB2, T25713;, TBP3.DAP

Organisms: Arabidopsis thaliana

Libraries: JASPAR 2024 [1], AthalianaCistrome v4_May2016 [2]

[1] Rauluseviciute I, Riudavets-Puig R, Blanc-Mathieu R, Castro-Mondragon JA, Ferenc K, Kumar V, Lemma RB, Lucas J, Cheneby J, Baranasic D, Khan A, Fornes O, Gundersen S, Johansen M, Hovig E, Lenhard B, Sandelin A, Wasserman WW, Parcy F, Mathelier A. JASPAR 2024: 20th anniversary of the open-access database of transcription factor binding profiles. Nucleic Acids Res : (2023). [Pubmed]

[2] O'Malley RC, Huang SS, Song L, Lewsey MG, Bartlett A, Nery JR, Galli M, Gallavotti A, Ecker JR. Cistrome and Epicistrome Features Shape the Regulatory DNA Landscape. Cell 165:1280-92 (2016). [Pubmed]

Notes: DAP-seq

Length: 19

Consensus: awwwhtwwr**CCCTAA**wtwy

Weblogo:

PSSM:

P0	A	C	G	T	
01	266	105	68	133	a
02	277	76	68	151	w
03	262	91	58	161	w
04	237	114	47	174	w
05	171	148	43	210	h
06	119	80	76	297	t
07	302	9	86	175	w
08	378	3	1	190	w
09	335	0	237	0	r
10	42	530	0	0	C
11	0	572	0	0	C
12	0	572	0	0	C
13	0	0	0	572	T
14	572	0	0	0	A
15	483	0	87	2	A
16	191	87	12	282	w
17	95	117	25	335	t
18	144	137	59	232	w
19	140	148	88	196	y

Type: **Heterodimer**

Binding TFs: T25713 (Myb-like DNA-binding domain, linker histone H1 and H5 family, Myb-like DNA-binding domain)
Q9JFW5
Q9JFW5

Fig. 4 FootprintDB landing page for motif TBP3:MA1355.1:JASPAR, accessible at https:/footprintdb.eead.csic.es/index.php?motif=TBP3:MA1355.1:JASPAR. Note that the sequence logo is also available in trimmed version in the JASPAR website

5. Repeat **steps 1–10** in Subheading 3.2 and save "query genes" and "sequences" locally using, respectively, "random.genes" and "random.fasta" tags (*see* **Note 10**).

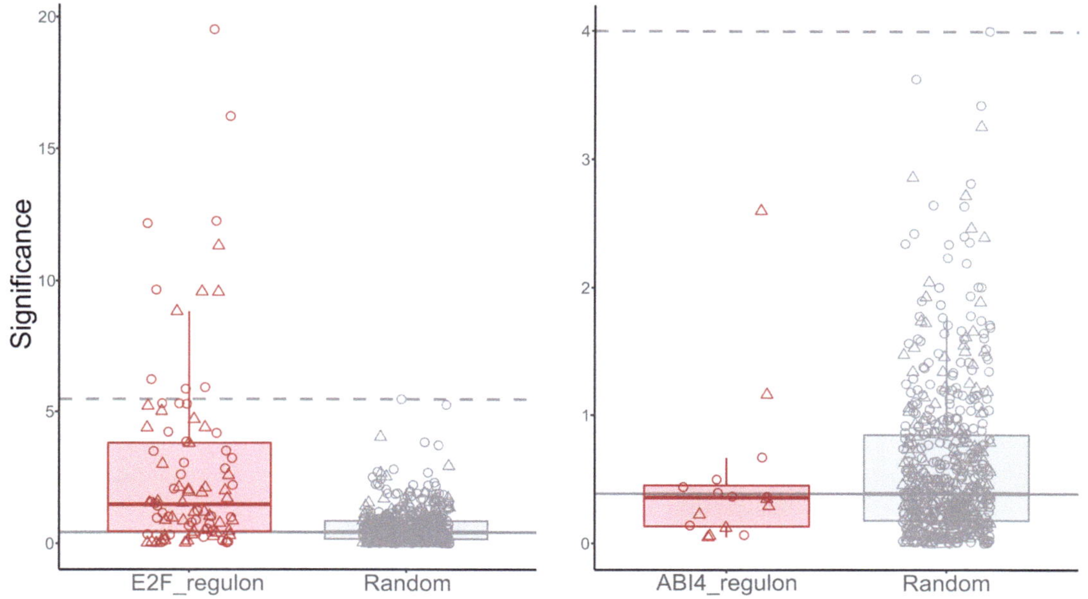

Fig. 5 Distributions of DNA motif significance, for example, clusters E2F (left) and ABI4 (right). For each cluster, two boxplots summarize the significance of motifs discovered on regulons and on random clusters of the same size with algorithms oligo-analysis (circles) and dyad-analysis (triangles). The y-axis shows the statistical significance $-\log_{10}$ (*P*-value) of all motifs discovered, with dashed lines marking the most significant motif found in negative controls

3.4 Interpretation of Results

The last stage of the protocol is the interpretation of results, which requires having at hand results of both clusters of co-expressed genes and random clusters, serving as negative controls.

3.4.1 Quality Control of Discovered Motifs

To assess the quality of any discovered motifs through this protocol, the comparison of identified motifs between the cluster of co-expressed genes and the random control sets is essential. Specifically, the statistical significance of those motifs needs to be compared. The distribution of motif significance in Fig. 5 illustrates this comparison, showcasing two example clusters from prior research [9]. The most significant motifs predicted by both *oligo-analysis* and *dyad-analysis* algorithms (red bars) are shown next to the significance of motifs discovered in random clusters (gray bars). The motif on the right (ABI4) exhibits lower significance compared to motifs found in random clusters of the same size, and therefore should not be considered as a reliable prediction. In contrast, the motif on the left (E2F) clearly exceeds the significance of those identified within random gene sets making it a promising candidate motif. Furthermore, our work has revealed that the number and significance of discovered motifs can vary depending on the size of the upstream window employed [8]. Therefore, trying all Up1 to Up4 windows is recommended (*see* **steps 2** and **4** from

Subheadings 3.2 and 3.3, respectively). Although the motifs are not shown in Fig. 5, a visual inspection is always a good practice; sometimes we can immediately identify motifs that look like artifacts or that may be repetitive elements, not necessarily TF binding motifs.

To ensure reproducibility, the code used to generate the plot depicted in Fig. 5 is accessible in Subheading 3.3 of the Docker tutorial (*see* **Note 10**). Note that a previous protocol [9] describes further ways of validating motifs enriched in clustered sequences provided and they are part of the current Ensembl Plants release.

3.4.2 Motif Occurrences Along Promoters

Based on our experience, we have often observed that TFs from specific families tend to exhibit a binding preference in certain bins within Up1 to Up4 windows (*see* for instance [8]). The positional distributions generated by peak-motifs (Fig. 3) provide valuable insights into these preferences.

Such distributions can also be produced using the Docker image available at https://hub.docker.com/r/biocontainers/rsat. When dealing with variable upstream region length, we recommend scanning the motif along the largest upstream stretch (referred to as Up1 in this tutorial) to gain a more comprehensive overview of its spatial distribution. *See* Subheading 3.4 of the Docker tutorial on **Note 10** to find code to produce such plots in R.

3.4.3 Analysis of Putative Binding TFs

Besides the discovered motifs, which are the primary output of this protocol, in previous experiments we have found it useful to analyze in parallel the TFs that might be binding to those motifs. There are two reasons for this: i) If a DNA motif explains the co-expression of a gene cluster, then we can expect that one or more TFs might be binding to sites upstream or downstream of those genes; ii) the DNA motifs annotated at curated collections provide independent estimates of the binding specificities of TFs, which can help validate or reject any discovered motifs. In the context of this protocol, there are several options you might want to consider:

1. Identify genes encoding TFs within co-expressed clusters. For this purpose, plant TF resources such as iTAK [15] or PlantTFDB [16] may be useful. Once these TF-encoding genes are shortlisted, examine their expression profiles across the conditions/tissues or interest, and retrieve their corresponding protein sequences.
2. Predict the putative DNA motifs of the TFs using footprinDB and following the guidelines outlined in the previous protocol [17]. Please note that while RSAT detects de novo motifs from clusters of co-expressed genes, instead footprintDB predicts putative motifs bound by individual TFs based on protein sequence similarity (see also https://github.com/eead-csic-compbio/footprintDBclient).

3. Compute the motif-to-motif similarity between cluster-enriched motifs (identified in **step 11**, Subheading 3.2) and TF-enriched motifs (identified in **step 2**, this section), using RSAT tool "compare matrices" and the similarity score "Ncor." High Ncor values indicate pairs of highly similar motifs, which can subsequently be used for functional annotation. To perform this comparison, navigate to "Matrix tools" on the left-hand menu, and then click on "compare matrices." Within the query matrices box, paste the footprintDB-predicted motif in transfac or tab format. In the Reference matrices (database or custom motif collection) box, upload the RSAT predicted matrix (**step 11**, Subheading 3.2) then click on "GO" option to launch the analysis.
4. To ensure comparative and reliable findings, you might want to generate a control subset of 50 random TFs along with their corresponding protein sequences. Ideally, this control subset should not contain genes found in the co-expressed gene cluster. In other words, using the example in Fig. 6, control subsets of random TFs for cluster M6 should be ideally selected from genes outside M6.
5. Perform **steps 2** and **3** for the random clusters, and plot the computed "Ncor" scores for better visualization of the results (Fig. 6).

For instance, Fig. 6 shows footprintDB-predicted motifs for peach transcription factors Prupe.5G106700 and Prupe.6G303500, which were grouped in clusters M6 and M7 and annotated, respectively, as WRKY and bHLH TFs in [8]. These motifs were compared to the significant motifs discovered upstream

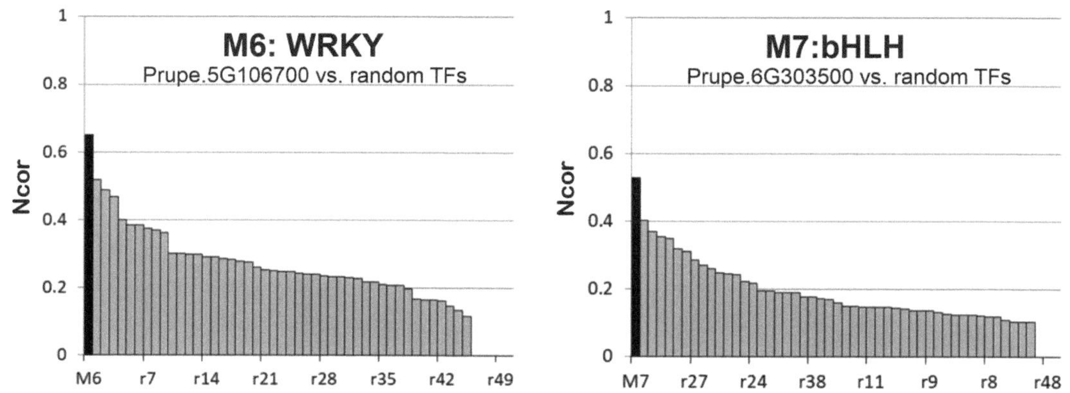

Fig. 6 Comparison of footprintDB-predicted motifs, inferred from protein sequence conservation to motifs enriched in the promoters of modules M6 and M7 in [8]. The y-axis shows Ncor values of the most significant motif predicted in a module (black bar) and the motifs predicted for 50 randomly sampled peach transcription factors (gray bars)

of genes in those modules (black bars). Instead, the gray bars summarize the comparisons of predicted motifs for 50 randomly sampled peach TFs. In other words, the examples in the figure evaluate how similar are motifs found upstream of co-expressed genes to the motifs predicted for TFs in the same clusters.

4 Notes

1. Gene models change across versions; make sure you choose the right assembly and annotation. If the genome of interest is not available at https://rsat.eead.csic.es/plants/supported-organisms.cgi, please contact us at rsat-contact@list01.bio.ens.psl.eu to request it. Four flat files are required for a new species (raw genomic sequence, repeat-hard-masked genomic sequence, gene annotation, and peptide sequences of CDS features; *see* https://rsa-tools.github.io/managing-RSAT).
2. In this protocol, small clusters are left out of the analysis, as in our experience the statistical approaches tested here require at least 10–15 sequences.
3. A crucial parameter to evaluate the results of motif discovery is to estimate the rate of false positives (FPs). RSAT programs compute a significance score, which is the minus log of the expected number of false positives. For example, a motif associated with a significance of 1 should be considered as poorly significant, since on average we would expect 0.1 FPs, i.e., one every ten random trials. In contrast, a significance of, e.g., 16 is very promising, since on average such a result would be expected every $10E^{-16}$ random trials. However, the theoretical significance relies on the correctness of the background model (computed here as k-mer and dyad frequencies in the whole set of promoters). In some cases, sets of plant promoters can discard from the theoretical model, due to heterogeneity of the input (e.g., inclusion of repetitive sequences). The negative control consists in measuring the significance obtained by submitting a random selection of promoters from the organism of interest. Although each of these genes is likely to be regulated by distinct TFs (and its promoter should contain corresponding binding sites), in principle the random set as a whole should not be co-regulated, so that the elements would differ from gene to gene, and there should thus be no overrepresented motif in their promoters.
4. The genomic delimitation of the proximal promoter region is far from being straightforward. Without prior knowledge of exact boundaries, analyzing promoters of variable length is a good alternative to reveal the most informative windows and to get a comprehensive picture of the regulatory context.

5. Should the connection to the server interrupt, it might be safer to go back and choose "email" as delivery option.
6. One gene ID per row, only the first word of each row is considered.
7. This program is generally relevant when analyzing sets containing a large number of sequences with positional bias, such as ChIP-seq peaks or genome-wide promoter sets, where we expect to find motifs at the center of the peaks and upstream the transcription start sites, respectively. Besides, position-analysis and local-word-analysis will not run if a control set is provided.
8. Option "*Origin*" indicates the reference position relative to each sequence (start, center, or end). When this option is set to "end," the coordinates are computed relative to the end of the sequence, with negative values indicating upstream location. Option "*Offset*" allows shifting the reference position by a given number.
9. Collections of plant DNA motifs are still very fragmentary with the exception of *Arabidopsis thaliana* (Table 2). Considering this, one might be tempted to explore more complete collections such as *footprintDB* or *JASPAR core all*. However, the results should be interpreted with caution, as *cis* regulation in plants and other kingdoms of the tree of life is not necessarily conserved.
10. For a large number of negative controls, this task can be tedious. The tutorial at https://eead-csic-compbio.github.io/coexpression_motif_discovery/peach/Tutorial.html explains how to automatize these steps and plot the results by using a Docker container.

Acknowledgments

We are grateful to all members of the RSAT team for continuous support and cooperation. This work was funded by the Spanish Ministry of Economy and Competitiveness (AGL2017-83358-R MCIU/AEI/FEDER/UE), the Government of Aragón (A09_23R and A08_23R, cofinanced with FEDER funds), and CSIC (FAS2022_052, INFRA24018). N.K. was funded by a 2-month Erasmus+ traineeship at Aix-Marseille University. N.K. and F.M.T hold PhD contracts awarded by the Government of Aragón. A.A. was funded by a Master's scholarship from CIHEAM Zaragoza.

Conflict of Interest The authors have no conflicts of interest to declare that are relevant to the content of this chapter.

References

1. Beer MA, Tavazoie S (2004) Predicting gene expression from sequence. Cell 117:185–198
2. Sand O, Turatsinze J-V, Van Helden J (2008) Evaluating the prediction of cis-acting regulatory elements in genome sequences. In: Frishman D, Valencia A (eds) Modern genome annotation. Springer, Vienna, pp 55–89
3. Koschmann J, Machens F, Becker M et al (2012) Integration of bioinformatics and synthetic promoters leads to the discovery of novel elicitor-responsive cis-regulatory sequences in arabidopsis. Plant Physiol 160:178–191
4. Yu C-P, Chen SC-C, Chang Y-M et al (2015) Transcriptome dynamics of developing maize leaves and genomewide prediction of cis elements and their cognate transcription factors. Proc Natl Acad Sci USA 112:E2477
5. Schmidt T (1998) Genomes, genes and junk: the large-scale organization of plant chromosomes. Trends Plant Sci 3:195–199
6. Abdalrahem A (2022) Exploring the panregulome of grasses with model species Brachypodium distachyon. https://digital.csic.es/handle/10261/284710
7. Contreras-Moreira B (2023) Statistics of plant genomes at RSAT::Plants. https://rsat.eead.csic.es/plants/data/stats
8. Ksouri N, Castro-Mondragón JA, Montardit-Tarda F et al (2021) Tuning promoter boundaries improves regulatory motif discovery in nonmodel plants: the peach example. Plant Physiol 185:1242–1258
9. Contreras-Moreira B, Castro-Mondragon JA, Rioualen C et al (2016) RSAT::Plants: Motif discovery within clusters of upstream sequences in plant genomes. Methods Mol Biol 1482:279–295
10. Sebastian A, Contreras-Moreira B (2014) footprintDB: a database of transcription factors with annotated cis elements and binding interfaces. Bioinformatics 30:258–265
11. Thomas-Chollier M, Darbo E, Herrmann C et al (2012) A complete workflow for the analysis of full-size ChIP-seq (and similar) data sets using peak-motifs. Nat Protoc 7:1551–1568
12. van Helden J, Andre B, Collado-Vides J (1998) Extracting regulatory sites from the upstream region of yeast genes by computational analysis of oligonucleotide frequencies. J Mol Biol 281: 827–842
13. Rauluseviciute I, Riudavets-Puig R, Blanc-Mathieu R et al (2024) JASPAR 2024: 20th anniversary of the open-access database of transcription factor binding profiles. Nucleic Acids Res 52:D174–D182
14. Turatsinze J-V, Thomas-Chollier M, Defrance M et al (2008) Using RSAT to scan genome sequences for transcription factor binding sites and cis-regulatory modules. Nat Protoc 3: 1578–1588
15. Zheng Y, Jiao C, Sun H et al (2016) iTAK: a program for genome-wide prediction and classification of plant transcription factors, transcriptional regulators, and protein kinases. Mol Plant 9:1667–1670
16. Tian F, Yang D-C, Meng Y-Q et al (2019) PlantRegMap: charting functional regulatory maps in plants. Nucleic Acids Res 1:1–10
17. Contreras-Moreira B, Sebastian A (2016) FootprintDB: analysis of plant cis-regulatory elements, transcription factors, and binding interfaces. In: Hehl R (ed) Plant synthetic promoters. Springer, New York, pp 259–277

Chapter 15

Inferring Gene Networks from High-Throughput Transcriptomes

David Navarro-Payá, Luis Orduña, José D. Fernández, Antonio Santiago, Elena A. Vidal, and José Tomás Matus

Abstract

Systems biology has helped to better understand the complexity of gene-to-gene interactions at a genome-wide level through the practical application of network theory. Over the last decade, different types of gene networks have been built to assist in gene function prediction and the modeling of transcriptional regulation in plants. While gene co-expression networks (GCNs) focus on general gene-to-gene relationships, gene regulatory networks (GRNs) depict the hierarchical connections among transcription factors and target genes. Here, we provide two strategies to produce gene networks based on high-throughput transcriptomic data—aggregated GCNs (aggGCNs) and inferred GRNs. A custom *in-house* pipeline is presented to build aggGCNs, while GRNs are inferred through the implementation of the GENIE3 algorithm. All of the presented code is publicly available, both in this chapter and associated Git repositories, and even though the workflows have been developed for particular plant species such as grapevine and tomato, they can be adapted to any other plant species or eukaryotes.

Key words Gene co-expression networks, Gene regulatory networks, Machine learning, Plant transcriptomic regulation, Genie3, AUROC

1 Introduction

Gene expression is a dynamic process where DNA is transcribed and then often translated to fulfil a function. Eukaryotic organisms mainly regulate spatiotemporal gene expression at epigenomic, transcriptional, and posttranscriptional levels. Among these processes, transcriptional regulation has been described as a key regulatory mechanism in plant species [1], in line with the observation that transcript levels generally correlate with protein abundance. In fact, transcriptional regulation in plants, given their sessile nature, is key in responses to both abiotic and biotic stresses. Therefore, grasping the complexity of transcriptional control is a key element

José M. Franco-Zorrilla and Carmen Martín-Pizarro (eds.), *Plant Transcription Factors: Methods and Protocols*, Methods in Molecular Biology, vol. 2985, https://doi.org/10.1007/978-1-0716-4972-5_15,

of future developments in plant science and agriculture as we better understand how plants adapt to their changing environment.

Transcriptional regulation is mainly exerted through the action of transcription factors (TFs), which are DNA-binding proteins that act in a coordinated manner to repress or promote gene expression. Land plants, and particularly angiosperms, have a high number and diversity of TFs as these have largely expanded in copy number within their genomes throughout evolution [2], and there is a correlation between the number of TFs present in the genome and organism complexity across the plant tree of life [3]. With the advent of systems biology approaches and greater computational resources, in silico strategies can now provide a holistic view of gene-to-gene interactions in an organism. Indeed, network science has become, over the last decade, a central paradigm in molecular systems biology as a robust modeling framework [4].

A gene network is made up of nodes and edges, corresponding to genes and their interactions, respectively. These can be broadly classified into two categories: gene co-expression networks (GCNs), which look at general gene-to-gene interactions, and gene regulatory networks (GRNs), which focus on TF-target relationships. In GCNs, edges represent a measure of co-expression. Following the "guilt-by-association" principle, co-expressed genes will likely be involved in the same function or pathway [5]. These are often context-dependent and dynamic, meaning that they can vary depending on the tissue type, developmental stage, environmental condition, or stress response, representing a further layer of complexity. Furthermore, biological networks are characterized by a scale-free topology [6], defined by the fact that the distribution of the nodes conforms to a power law, which means that most of the nodes will have a low number of connections, while a few will be highly connected (Fig. 1). Another key aspect in network theory is the study of the overall and local topology of a network, for which there are a range of established metrics (e.g., "degree," "node transitivity," "betweenness," and "eccentricity"), all of which are suitable to better identify and understand the role of relevant molecular components in a particular biological system. For instance, a highly connected node can be referred to as a hub, and these are particularly interesting as they can play key roles in integrating biological processes. It's generally accepted that TFs act many times as hubs within these networks.

The use of GCNs arises as a reliable and data-driven gene function prediction tool. Advances in genomics have enabled the exploitation of GCNs through different approaches to predict the function of uncharacterized genes in different plant model species [7, 8]. Among the plethora of available methods, it has been shown that the highest reciprocal ranking (HRR) [9] of Pearson's correlation values can be applied to build reliable global networks [10]. In addition, a few studies carried out in model species (e.g., in mouse

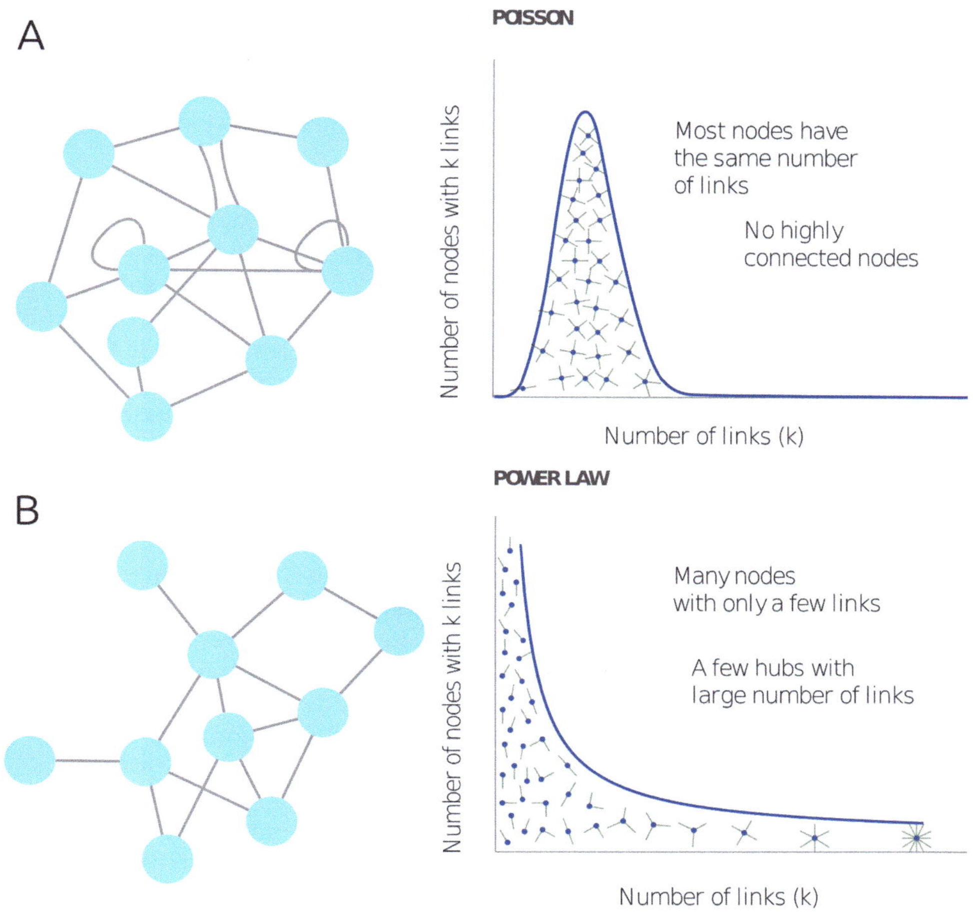

Fig. 1 Gene co-expression and regulatory networks follow a power law distribution, which leads to a scale-free topology. (**a**) Random network with a Poisson distribution of nodes' degree, showing that nodes have a similar number of links. (**b**) Scale-free topology network, showing that nodes' degree distribution fits a power law curve where most of the genes are poorly connected, and a few nodes are largely connected, which can be referred to as hubs (Adapted from [6])

and yeast, tomato, maize, and *Arabidopsis*) have shown that aggregation of individually and independently generated GCNs [11, 12] can improve overall gene function prediction. Several studies have shown the potential of GCNs built with transcriptomic expression datasets (microarray or RNA-seq) as in silico gene function prediction tools in different model species using different approaches [7, 8, 13].

Moreover, the study of TF-target regulation can be pursued by modeling GRNs. The GENIE3 (gene network inference with ensemble of trees) algorithm is an efficient method to obtain inferred GRNs [14, 15]. GENIE3 generates direct TF-target gene pairings from high-throughput gene expression data using a tree-based iterative approach. GENIE3 does not rely on labeled examples of gene regulatory interactions, but it does require a list of an organism's TFs. The method identifies ranked regulatory pairings based solely on the structure of expression data. Furthermore, GENIE3 has proved useful in predicting in vivo regulation across a variety of plant species [16–21].

The present chapter is meant as a guide into two complementary network construction methods (aggregated GCNs and GENIE3 GRNs). Both of these can be useful tools in gene function prediction, and although a set of hardware and software requirements must be met, this chapter aims to serve as a step-by-step tutorial on building these gene networks. The code for both pipelines is freely available and can be adapted to any species of interest with a sufficiently large number of expression datasets available.

2 Materials

The two complementary approaches to building networks, for GCNs or GRNs, presented here have both shared and specific software and hardware requirements. The version of required software and packages has always been indicated for reproducibility. Although more recent versions of the software could still work, we cannot guarantee correct or full functionality. Consider using docker [22], or a similar container platform, with the indicated versions for exact reproducibility.

The system requirements will vary according to the amount of data to be evaluated. The hardware requirements can be particularly demanding when considering the larger volumes of data for GCN generation, and even though a high-performance computer cluster is also recommended for GRNs, it may be possible to use a more typical computer setup provided the input data is more modest in size. Nevertheless, a high-performance computer (HPC) with a Linux operative system is highly recommended in both cases.

2.1 Aggregated GCNs

Building aggregated gene co-expression networks requires the following:

- Python3 (preinstalled in Linux based distributions) (v3.8).
- R (v4.2.1) and the following R packages:
  - Optparse (v1.7.4).
  - data.table (v.1.15.4),
  - parallel (v4.2.1),
  - igraph (v2.0.3),
  - DESeq2 (v1.38.3).
- fastp (v0.21.0).
- STAR (v2.7.3a).

- subread (v2.0.0).
  - featureCounts (v2.0.0),
- SRA toolkit (v2.11.2).
  - prefetch (v2.11.2),
  - fastqdump (v2.11.2),

2.2 GRNs

Building inferred gene regulatory networks require the following:

- R (v4.1) and the following R packages:
  - data.table (v1.15.4),
  - GENIE3 (v1.22.0) [14].
  - dplyr (v1.1.4),

2.3 Network Parameter Analysis

- R (v4.1) and the following R packages:
  - igraph (v1.5.1),

2.4 Network Visualization

- R (v4.1) and the following R packages:
  - networkD3 (v0.4).
  - Influential (v2.2.9) [23].

3 Methods

3.1 Aggregated Gene Co-expression Networks

The aim of this section is to guide the user in building aggregated gene co-expression networks (aggGCNs) from scratch, provided the above hardware and software requirements are met. A basic understanding of Bash and Python is also recommended. All the code used in this section is fully available as two public repositories—raw_counts_generation (https://github.com/Tomsbiolab/raw_counts_generation) and agg_WGCN (https://github.com/Tomsbiolab/agg_WGCN)—which are currently adapted to work with grapevine (*Vitis vinifera*). Nevertheless, they can be adapted to work with a different plant species or indeed any other eukaryote. Moreover, they are compatible with SLURM managed servers, but the master bash scripts would also work in standard Linux systems.

Cloning the repository after fulfilling hardware and software requirements mentioned in Subheading 2.1 should be sufficient for successfully generating an aggregated gene co-expression network based on input data of choice, i.e., a list of SRA (Sequence Read Archive) run IDs. The pipeline is optimized to work with SRA data available at National Center for Biotechnology Information (NCBI) servers. Per-run metadata is readily accessible through NCBI, although download of bulk metadata is advisable through the inbuilt Run Selector. Careful curation of metadata is strongly recommended if a condition-dependent network is to be generated. Since aggregated networks benefit from larger transcriptomic datasets, a minimum of 20 SRA bioprojects is recommended, each containing at least 4 SRA runs, for optimal network performance [24].

3.1.1 Input Data Preprocessing

This section corresponds to the "raw_counts_generation" Git repository, which takes a list of SRA IDs as input and returns individual raw count matrices (*see* **Note 1**). Although the necessary steps are automated within the code, a general outline is provided below:

1. Download of SRA runs using prefetch:

```
prefetch "$runid" --verify yes --check-all -X 100G -o
"$runid".sra
```

2. Processing sra-formatted files into fastq files using fastq-dump:

```
fastq-dump -I --split-files "$exp".sra -O "$exp"
```

3. Quality read trimming with fastp (single-end example):

```
fastp -j "$exp"_fastp.json -h "$exp"_fastp.html -w $cpus
n_base_limit 5 cut_front_window_size 1 cut_front_mean_quality
30 --cut_front cut_tail_window_size 1 cut_tail_mean_quality
30 --cut_tail -l 20 -i "$exp"_1.fastq -o "$exp"_1_trim.fq.gz
```

4. Trimmed reads aligned to an indexed genome using STAR [25] (single-end example):

```
STAR --runMode alignReads --genomeDir "$indexed_genome_file"
--runThreadN $cpus --readFilesCommand zcat --readFilesIn "$ex-
p"_1_trim.fq.gz --outSAMunmapped Within --outSAMtype BAM Un-
sorted --limitOutSJcollapsed 8000000 --limitIObufferSize
220000000
```

5. Sorting of aligned reads using samtools to fulfil featureCounts requirements:

```
Samtools sort -@ $cpus Aligned.out.bam -o "sorted_"$exp".bam"
```

6. Raw count matrix generation with FeatureCounts [26]:

```
featureCounts -p -T 20 -t "gene" --minOverlap 1 -C -a
"$annotation_file" -o count_matrices/"$exp"_count_matrix.txt
"sorted_"$exp".bam"
```

3.1.2 Network Generation

This section corresponds to the "agg_GCN" Git repository that uses Subheading 3.1.1 raw count matrices as input, while it outputs a final aggregated gene co-expression network (aggGCN) (*see* **Note 2**). Although the necessary steps are automated within the code, a general outline is provided below:

1. Count matrices are grouped according to bioproject. Runs that have less than ten million read alignments are discarded by default (*see* **Note 3**).
2. Bioprojects with less than four valid SRA runs are also discarded (*see* **Notes 4** and **5**).
3. Generation of a raw count matrix per Bioproject ("loop_merge_matrix.py").
4. Normalization of expression to FPKM (fragments per kilobase per million reads) for each Bioproject. Lowly expressed genes (less than 0.5 FPKM across all runs of Bioproject) will be removed.
5. Generation of a Pearson's correlation coefficient (PCC) matrix for expressed genes per Bioproject ("loop_PCC.py") (*see* **Note 6**).
6. Generation of binary highest reciprocal ranking (HRR) matrices. For each Bioproject, its PCC matrix will be analyzed as follows: For each gene present in the matrix, the rest of the genes will be sorted in descending order according to their PCC value with respect to the analyzed gene. Then, the top 1% PCC ranking genes will be given a value of 1 in the PCC matrix, and the rest of values will be changed to 0. Once this process has been applied to all the genes present in the PCC matrix, ranked PCC values will be used to compute each HRR matrix [9]. The HRR matrix is generated by applying the following formula to each possible pair of genes:

$$HRR(A,B) = max(rank(A,B), rank(B,A))$$

Since the ranked PCC matrix only contains digits "0" and "1," the generated HRR matrix is a binary and symmetric matrix ("loop_HRR_generation_TOP420.py").

7. Co-occurrence computation. For each possible pair of genes, the co-occurrence value is the number of HRR matrices (one per bioproject) in which the HRR value for the given pair is equal to 1 ("co_occurrence_data.table.R") (*see* **Note 7**).
8. Filtering co-occurrence matrix. The matrix generated in step 7 is filtered by keeping only the top 1% strongest interactions for each gene in which the gene is source (*see* **Notes 8** and **9**). The network output is both in a tabular format for Cytoscape [27], a common network visualization software, and as an EGAD [28] compatible matrix format which serves to evaluate network performance.

3.2 Gene Regulatory Networks

The following sections describe the pipeline for generating an inferred gene regulatory network (GRN) with the GENIE3 algorithm [14]. The following section is meant as a tutorial on running GENIE3 algorithm to predict regulatory interactions and their potential hierarchical structure, with a particular focus on transcription factors and their target genes. In addition to meeting hardware and software requirements above, a basic understanding of R and bash is recommended. Wherever R code is provided in the examples below, a new line is indicated by the ">" symbol:

```
# Example R command
> print("Executing R script...")
```

The pipeline consists of the following steps:

- Compilation of transcriptomes: The RNA-seq counts must be collected into a counts table, formatted correctly to allow the algorithm to detect the processed reads for each SRA libraries.
- Applying the GENIE3 algorithm: After meeting software and hardware requirements, run the pipeline to generate a list of possible regulatory interactions between transcription factors (TFs) and their target genes.
- Obtaining inferred gene regulatory networks: Once GENIE3 calculations are completed, the output will include a comprehensive list of TF-target pairings. To improve the output GRN, a filter selecting the highest-scoring regulatory pairs should be implemented.

As with aggregated gene co-expression networks, the transcriptomic input selected by the user will define the type of final network obtained. For instance, choosing specific transcriptomes corresponding to a particular tissue will result in a condition-dependent network. There are no strict requirements in terms of minimum input transcriptome size for GENIE3; however, the general guideline is to have as many samples as possible, as recommended for multiple gene network inference methods [12, 15]. Users who follow these recommendations can construct GRN models using the GENIE3 algorithm to discover regulatory interactions in high-throughput transcriptomic data. Although the examples provided focus on tomato transcriptomes, this procedure is applicable to any plant species with sufficient transcriptomic data available.

3.2.1 Applying the GENIE3 Algorithm

The algorithm requires two main inputs:

- Transcriptome count data can be generated as described in Subheading 3.1.1, although it must be combined into one matrix (*see* **Note 10**). Note that GENIE3 does not require the data to be normalized, allowing the use of raw counts directly.

– A genome-wide list of gene IDs that are identified as TFs, which will represent the source nodes in the inferred GRN. Curated lists of TFs can typically be found in online databases grounded on structural genome annotation, such as TAIR [29], PlantTFDB [30], etc. Additionally, users can generate their own list of TFs by analyzing the gene functions based on alignments, genome annotations, orthologous TFs from other species, etc. (*see* **Note 11**).

The script below is an example use of GENIE3 to generate a GRN from organ-specific transcriptome datasets and a curated list of tomato TFs.

```
#Loading libraries
>library(dplyr)
>library(GENIE3)

#Setting working directory
>setwd("/home/username/working_folder/")

#Loading the datasets
>dseq_sig <- read.table("tissue1_counts.txt", header=T,
sep="\t")
>reg <- read.table("curated_tfs.txt", header=F)

>rownames(dseq_sig)<-dseq_sig$Gene
>dseq_sig$Gen <- NULL
>exprMatr <- data.matrix(dseq_sig)

#Setting a seed for reproducibility
>set.seed(123)

#Running GENIE3
>weightMat <- GENIE3(exprMatr, regulators = reg$V1, nCores =
40, verbose = T)
>Genie3_file <- getLinkList(weightMat, reportMax = NULL)
>write.table(Genie3_file, "inferred_grn_tissue1.txt")
```

3.2.2 Obtaining Gene Regulatory Networks

The raw GENIE3 GRN output generally includes millions of predicted regulatory interactions organized into four columns: "row_id," "TF," "Target," and "row_weight," the latter providing a ranked score marking the likelihood of a regulatory interaction [15]. A higher TF-target pairing indicates a greater probability of a genuine regulatory interaction. Considering that the raw output contains many putative interactions with low scores that may represent false positives, a noise-filtering step is recommended. Keeping the top 2% or top million scored interactions is typically recommended [19, 31] to select the most likely biologically relevant interactions. However, the filtering threshold may be adjusted based on specific conditions studied. The following R script provides an example of such filtering:

```
# Load required libraries
library(dplyr)
library(data.table)

# Set working directory
>setwd("/home/username/working_folder/")

# Read the data file
>Genie3_net <- fread("inferred_grn_tissue1.txt", header = T,
sep = " ", quote = "\"")

# Rename columns
>colnames(Genie3_net) <- c("row_id", "TF", "TARGET",
"weight")

# Sort the data by TARGET and weight in descending order
>genie3.table <- Genie3_net %>% arrange(desc(TARGET), desc
(weight))

# Calculate percentage for each group of TARGET
>sort_percent <- genie3.table %>% group_by(TARGET) %>%
mutate(percent = round(row_number() * 100 / n(), 2))

# Filter data for the top 1% of each TARGET group based on
weight
>filt_sort_per <- sort_percent %>% filter(percent <= 2)

# Write the filtered and sorted data to a new file
>write.table(filt_sort_per, "filtered_grn_tissue1.txt", sep
= "\t", row.names = F, quote = F)
```

The filtered subnetwork can be considered a final GRN model, holding a catalog of high-confidence TF-target pairings, which can be used to study an organism's regulatory pathways. Moreover, the network might be utilized to provide insights into specific stimuli or experimental conditions by narrowing the target list to represent regulation under specific stimuli or experimental conditions, increasing the network's ability to examine biologically relevant hypotheses. Indeed, condition-dependent GRNs from different tissues, time frames, treatments, or mutants, etc. can be used to reveal potential alterations in the regulatory pathways within the organism under study [18, 19].

Note that both aggGCNs and GRNs can benefit from overlapping with experimental regulatory data (*see* **Note 12**), since the results of both networks are computational models derived from gene co-expression data. By filtering predicted interactions to retain pairs supported by multiple experimental datasets, the resulting networks may have improved potential to closely represent the in vivo regulatory pathways occurring in plant cells [16, 17].

3.3 Analysis of Network Parameters

Gene networks, including GCNs and GRNs, can be systematically analyzed with the purpose to reveal key nodes with potential biological significance, especially in the context of regulatory cascades [23]. These analyses frequently focus on metrics derived from general network theory, such as centrality, neighborhood connectivity, outdegree, and clustering coefficients, among others. These can be used to identify nodes that could be acting as key in vivo regulators; for instance, the highest central genes often correspond to key players in biological processes. In this section, we provide a pipeline to perform network analysis and determine potential key hubs on GCNs and GRNs.

```
# Load required packages
library(igraph)
library(influential)

# Set working directory
>setwd("/home/username/work_folder")

# Read the network file
>edge_matrix <- fread("filtered_grn_tissue1.txt", header = T,
sep = " ", quote = "\"")

# 3. Construct the network
>network <- graph_from_data_frame(edge_matrix, directed = T)

# 4. Calculate influential node using IVI scores
>ivi_scores <- ivi(graph = network, directed = T)

# 5. Calculate network analysis metrics
>degree_vals <- degree(network)
>betweenness_vals <- betweenness(network)
>closeness_vals <- closeness(network)
>eigenvector_vals <- eigen_centrality(network)$vector
>outdegree_vals <- degree(network, mode = "out")
>neighborhood_co_vals <- neighborhood.connectivity(graph =
network, mode = "all")
```

```
# 6. Create a results table
>results <- data.frame(
Gene = V(network)$name,
Degree = degree_vals,
BetweennessCentrality = betweenness_vals,
ClosenessCentrality = closeness_vals,
EigenvectorCentrality = eigenvector_vals,
Outdegree = outdegree_vals,
NeighborhoodConnectivity = neighborhood_co_vals,
IVI = ivi_scores)

# 7. Write the results as a tab-separated text file
>output_file <- "analyzed_network_pGRN-tissue1.txt"
>write.table(results, file = output_file, sep = "\t", quote
= F, row.names = F, col.names = T)
```

3.4 Network Visualization

There are many tools available for network visualization, and most of them involve some sort of subsampling of the network. It is important to bear in mind that the typical number of edges and nodes in a whole-genome GCN makes full visualization impractical. Instead, a subset of edges and nodes must be selected for feasible visualization. One of the most used programs is Cytoscape [27], which we also recommend. Nonetheless, it is still interesting to explore other options such as the NetworkD3 R package, which displays a network dynamically, where, by default, the nodes repel and attract each other in a simulated 3D space based on the number and weight of edges between them. This results in a 3D visualization of a network where visual topology is a result of the interconnectedness of its nodes. This evaluation of a network is an interesting way to test hypotheses (Fig. 2).

4 Notes

1. The "raw_counts_generation" pipeline is controlled by the master script named "slurm_sra_to_counts_8.sh." The following parameters should be set by the user: (1) *v_genome*: genome assembly fasta file that is for alignment (indicating file extension); (2) *v_annotation*: annotation file (without indicating file extension); (3) *genome_folder*: path to genome annotation and assembly files; (4) scripts_folder: path to scripts; (5) *metadata_file*: path to metadata file; and (6) *results_folder*: path to output folder.
2. The "agg_GCN" pipeline is controlled by the master script named "slurm_aggregated_network.sh."

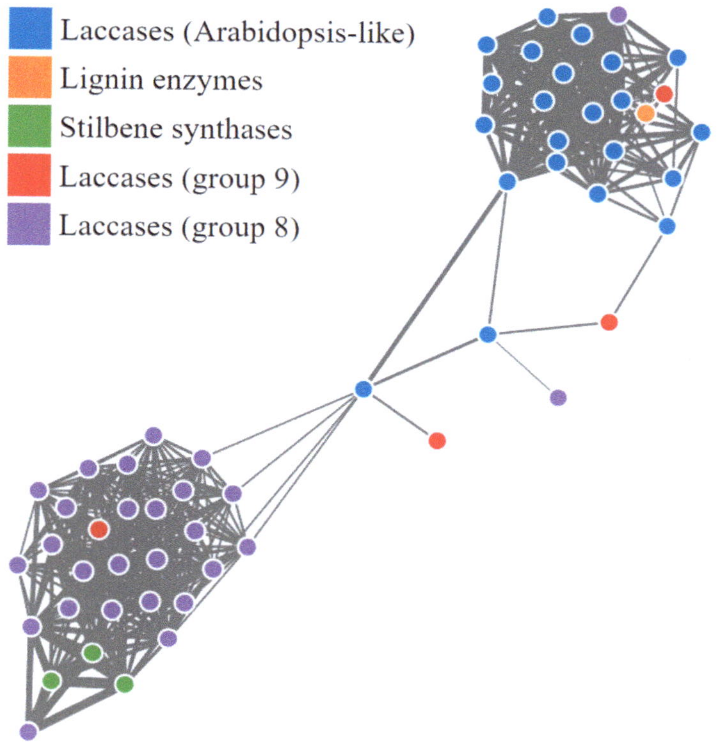

Fig. 2 Grapevine laccases cluster according to specific phenylpropanoid branch enzymes, suggesting neofunctionalization. The phenylpropanoid pathway has several branches leading to different specialized metabolites, such as anthocyanins, lignins, and stilbenes. Laccases in *Arabidopsis* are typically associated to lignin polymerization. A phylogenetic analysis has been used to color the nodes of different laccases in grapevine based on a close relationship with *Arabidopsis* laccases or whether they are part of the grapevine specific phylogenetic branches (groups 8 and 9). It appears that group 8 laccases are specifically co-expressed with stilbene synthase genes, suggesting neofunctionalization in the laccase lineage, perhaps having a role in the production of stilbene polymers known as viniferins. These findings have been previously published [32], and they highlight how network visualization can be used to examine a biological hypothesis

3. This filtering is performed to ensure that the libraries included in the GCN contain a minimal quantity of gene expression information. The filtering of libraries according to the number of alignments they contain can be performed using the information available in the summary files generated in **step 4**.
4. This filtering is performed because experiments with a reduced number of libraries will not generate robust PCC matrices, possibly adding noise to the final GCN. Nevertheless, this filtering step can be modified by editing the code, if needed (see line 22 of "preparing_experiments_folders.py").

5. A fully descriptive help is available for all Python scripts mentioned from this point onward. Help is accessible by typing *python3 script_name.py –help*.
6. Python script "loop_PCC.py" executes underneath the Rscript "generating_PCC.R," which uses the *cor()* R function to compute the PCC matrices.
7. The co-occurrence value of gene A to gene B is always going to be equal to the co-occurrence value of gene B to gene A.
8. Despite what was mentioned in **Note 7**, it is possible that a co-occurrence value between gene A and gene B is within the top 1% of gene A, but not in the top 1% of gene B. Therefore, the final aggregated GCN will only reflect the co-expression between gene A to gene B, but not gene B to gene A. Note that the edge between A and B will always be present in the full network even after the top 1% filtering of the co-occurrence matrix.
9. When filtering co-occurrence values, it is possible that, for a given source gene, the lowest co-expression value that makes the 1% cut will also be shared by other gene-pairs immediately below the table. In other words, there can be "ties" in co-expression values between gene-pairs co-expressed that passed or did not reach the threshold for a given source gene. The final filtered network would, in this case, always have the same number of source entries (corresponding to 1% of the genes in the genome) for a given gene. Indeed, the "top1_co_occurrence_matrix_version2_TOP420_removing_ties.py" script follows this approach. However, an alternative approach, including all gene-pairs that match the lowest co-expression value but still missed the threshold, will be added to the filtered-in entries. In this case, each source gene can have a different number of co-expressions (see "top1_co_occurrence_matrix_version2_TOP420_without_removing_ties.py").
10. To guarantee the correct function of the algorithm, the input data table should be organized as follows: The first column must be identified as "Gene" and should contain all of unique IDs corresponding to the genome annotation. The read counts determined for each gene must be included in the following columns, with each column representing a separate library obtained from the transcriptome runs.
11. The list of TF genes should comprise the known regulators of the organism under study. Where curated lists are not readily available, we recommend to access different databases such as PlantTFDB [30] in search for TF lists of closely related species and obtaining the desired TF list through orthology analysis. In addition, an InterProScan analysis [33] can be run as well on a protein multifasta input for de novo functional annotation

from which TFs can be identified. TF lists can also be filtered for those TFs expressed in an organ of interest, in case a condition-dependent inferred network is to be created.

12. Inferred GRNs described in this chapter can be improved by incorporating experimental data, such as DAP-seq, Y1H, and ChIP-seq, into the regulatory network model. Additionally, exploring differentially expressed genes (DEGs) derived from transcriptome analyses, in which a certain treatment is used or a TF is knocked down or overexpressed, can provide additional insight into the network targets' relevance under specific conditions. Finally, the GRN model can be further evaluated by performing gene set enrichment analyses on a TF's targets in a GRN, whose regulation of at least some targets has already been validated experimentally.

Acknowledgments

This work was supported by Grant PID2021-128865NB-I00 from the Ministerio de Ciencia, Innovación y Universidades (MCIU, Spain), Agencia Estatal de Investigación (AEI, Spain), and Fondo Europeo de Desarrollo Regional (FEDER, European Union) and the PROMETEO grant ENIGmA (PROMETEO/2021/056) and the associated PhD contract (PROMETEO/2021/056-01) awarded to AS by the Generalitat Valenciana (GVA). EV and JDF are supported by ANID-FONDECYT 1211130, ANID-Millennium Science Initiative Program ICN17_022, FOVI230159, and PhD scholarship ANID 21230478.

Conflict of Interest The authors have no conflicts of interest to declare that are relevant to the content of this chapter.

References

1. Yanagisawa S (1998) Transcription factors in plants: physiological functions and regulation of expression. J Plant Res 111:363–371. https://doi.org/10.1007/BF02507800
2. Fang Y, Jiang J, Hou X et al (2022) Plant protein-coding gene families: their origin and evolution. Front Plant Sci 13:995746. https://doi.org/10.3389/fpls.2022.995746
3. Yruela I, Oldfield CJ, Niklas KJ, Dunker AK (2017) Evidence for a strong correlation between transcription factor protein disorder and organismic complexity. Genome Biol Evol 9:1248–1265. https://doi.org/10.1093/gbe/evx073
4. Li Y, Pearl SA, Jackson SA (2015) Gene networks in plant biology: approaches in reconstruction and analysis. Trends Plant Sci 20: 664–675. https://doi.org/10.1016/j.tplants.2015.06.013
5. Wolfe CJ, Kohane IS, Butte AJ (2005) Systematic survey reveals general applicability of "guilt-by-association" within gene coexpression networks. BMC Bioinf. 6:227. https://doi.org/10.1186/1471-2105-6-227
6. Barabási A-L, Oltvai ZN (2004) Network biology: understanding the cell's functional organization. Nat Rev Genet 5:101–113. https://doi.org/10.1038/nrg1272

7. Burks DJ, Sengupta S, De R et al (2022) The Arabidopsis gene co-expression network. Plant Direct 6:e396. https://doi.org/10.1002/pld3.396
8. Madhamshettiwar PB, Maetschke SR, Davis MJ et al (2012) Gene regulatory network inference: evaluation and application to ovarian cancer allows the prioritization of drug targets. Genome Med 4:41. https://doi.org/10.1186/gm340
9. Mutwil M, Usadel B, Schütte M et al (2010) Assembly of an interactive correlation network for the Arabidopsis genome using a novel heuristic clustering algorithm. Plant Physiol 152:29–43. https://doi.org/10.1104/pp.109.145318
10. Liesecke F, Daudu D, Dugé de Bernonville R et al (2018) Ranking genome-wide correlation measurements improves microarray and RNA-seq based global and targeted co-expression networks. Sci Rep 8:10885. https://doi.org/10.1038/s41598-018-29077-3
11. Liesecke F, De Craene J-O, Besseau S et al (2019) Improved gene co-expression network quality through expression dataset downsampling and network aggregation. Sci Rep 9:14431. https://doi.org/10.1038/s41598-019-50885-8
12. Verleyen W, Ballouz S, Gillis J (2015) Measuring the wisdom of the crowds in network-based gene function inference. Bioinformatics 31:745–752. https://doi.org/10.1093/bioinformatics/btu715
13. Orduña L, Santiago A, Navarro-Payá D et al (2023) Aggregated gene co-expression networks predict transcription factor regulatory landscapes in grapevine. J Exp Bot 74:6522–6540. https://doi.org/10.1093/jxb/erad344
14. Huynh-Thu VA, Irrthum A, Wehenkel L, Geurts P (2010) Inferring regulatory networks from expression data using tree-based methods. PLoS One 5:e12776. https://doi.org/10.1371/journal.pone.0012776
15. Huynh-Thu VA, Geurts P (2019) Unsupervised gene network inference with decision trees and random forests. Methods Mol Biol 1883:195–215. https://doi.org/10.1007/978-1-4939-8882-2_8
16. Chen Y, Guo Y, Guan P et al (2023) A wheat integrative regulatory network from large-scale complementary functional datasets enables trait-associated gene discovery for crop improvement. Mol Plant 16:393–414. https://doi.org/10.1016/j.molp.2022.12.019
17. De Clercq I, Van de Velde J, Luo X et al (2021) Integrative inference of transcriptional networks in Arabidopsis yields novel ROS signalling regulators. Nat Plants 7:500–513. https://doi.org/10.1038/s41477-021-00894-1
18. Harrington SA, Backhaus AE, Singh A et al (2020) The wheat GENIE3 network provides biologically-relevant information in polyploid wheat. G3 (Bethesda) 10:3675–3686. https://doi.org/10.1534/g3.120.401436
19. Huang J, Zheng J, Yuan H, McGinnis K (2018) Distinct tissue-specific transcriptional regulation revealed by gene regulatory networks in maize. BMC Plant Biol 18:111. https://doi.org/10.1186/s12870-018-1329-y
20. Ranjan R, Srijan S, Balekuttira S et al (2024) Organ-delimited gene regulatory networks provide high accuracy in candidate transcription factor selection across diverse processes. Proc Natl Acad Sci USA 121:e2322751121. https://doi.org/10.1073/pnas.2322751121
21. Tu X, Mejía-Guerra MK, Valdes Franco JA et al (2020) Reconstructing the maize leaf regulatory network using ChIP-seq data of 104 transcription factors. Nat Commun 11:1–13. https://doi.org/10.1038/s41467-020-18832-8
22. Merkel D (2014) Docker: lightweight Linux containers for consistent development and deployment. Linux J 2014:76–91
23. Salavaty A, Ramialison M, Currie PD (2020) Integrated value of influence: an integrative method for the identification of the Most influential nodes within networks. Patterns (New York, NY) 1:100052. https://doi.org/10.1016/j.patter.2020.100052
24. Orduña-Rubio L, Santiago A, Navarro-Payá D et al (2023) Aggregated gene co-expression networks for predicting transcription factor regulatory landscapes in a non-model plant species bioRxiv:2023.04.24.538042. https://doi.org/10.1101/2023.04.24.538042
25. Dobin A, Davis CA, Schlesinger F et al (2013) STAR: ultrafast universal RNA-seq aligner. Bioinformatics 29:15–21. https://doi.org/10.1093/bioinformatics/bts635
26. Liao Y, Smyth GK, Shi W (2014) featureCounts: an efficient general purpose program for assigning sequence reads to genomic features. Bioinformatics 30:923–930. https://doi.org/10.1093/bioinformatics/btt656
27. Shannon P, Markiel A, Ozier O et al (2003) Cytoscape: a software environment for integrated models of biomolecular interaction networks. Genome Res 13:2498–2504. https://doi.org/10.1101/gr.1239303

28. Ballouz S, Weber M, Pavlidis P, Gillis J (2017) EGAD: ultra-fast functional analysis of gene networks. Bioinformatics 33:612–614. https://doi.org/10.1093/bioinformatics/btw695
29. Berardini TZ, Reiser L, Li D et al (2015) The Arabidopsis information resource: making and mining the "gold standard" annotated reference plant genome. Genesis 53:474–485. https://doi.org/10.1002/dvg.22877
30. Jin J, Tian F, Yang D-C et al (2017) PlantTFDB 4.0: toward a central hub for transcription factors and regulatory interactions in plants. Nucleic Acids Res 45:D1040–D1045. https://doi.org/10.1093/nar/gkw982
31. Olivares-Yañez C, Sánchez E, Pérez-Lara G et al (2021) A comprehensive transcription factor and DNA-binding motif resource for the construction of gene regulatory networks in Botrytis cinerea and Trichoderma atroviride. Comput Struct Biotechnol J 19:6212–6228. https://doi.org/10.1016/j.csbj.2021.11.012
32. Pilati S, Malacarne G, Navarro-Payá D et al (2021) Vitis OneGenE: a causality-based approach to generate gene networks in Vitis vinifera sheds light on the laccase and dirigent gene families. Biomolecules 11. https://doi.org/10.3390/biom11121744
33. Jones P, Binns D, Chang H-Y et al (2014) InterProScan 5: genome-scale protein function classification. Bioinformatics 30:1236–1240. https://doi.org/10.1093/bioinformatics/btu031

Chapter 16

Comparative Genomics Using *GeneDiscoveR*: A Toolbox for the Identification of Gene Candidates Associated with Plant Traits

Atilio O. Rausch, Juan Ignacio Tolopka, Tamara M. Ledesma, Linzhou Li, Sibo Wang, Shanshan Dong, Yang Liu, Agustín L. Arce, and Javier E. Moreno

Abstract

GeneDiscoveR, a novel R package, facilitates gene discovery in plant traits via comparative genomics. Despite the advancements in plant genome sequencing technologies, gene discovery in model and even more in non-model plants remains challenging. To address this gap, we introduce *GeneDiscoveR*, which enables the identification of orthogroups linked to specific plant traits or treatment responses. Leveraging extensive genomic data from diverse plant lineages, for instance, liverworts, we showcase its efficacy in identifying trait-specific genes. *OrthoFinder* defines orthologs, while *GeneDiscoveR* statistically detects trait-associated orthogroups. Here, we applied *GeneDiscoveR* to liverwort genomes to find enriched orthogroups in species with oil bodies within specialized cells or with many oil bodies in all cells. Additionally, we used it to identify OGs related to self-incompatibility from Brassicaceae genomes. This bioinformatics pipeline offers insights into plant trait genetics, aiding future gene discovery endeavors.

Key words Comparative genomics, Liverworts, Mapping, Gene orthology

1 Introduction

Gene discovery is critical for the future of plant improvement efforts aimed at enhancing yield, resilience, and nutritional value for a new green revolution. Despite tremendous advances in plant genome sequencing, there are still profound difficulties during the process of gene discovery, particularly for non-model plants that bear interesting and unique traits.

Large genome projects are being developed on plant species spanning a wide evolutionary range, with a strong overrepresentation of economically relevant plants. These sequences are shedding light on novel aspects of plant genomes, evolutionary history, and

José M. Franco-Zorrilla and Carmen Martín-Pizarro (eds.), *Plant Transcription Factors: Methods and Protocols*, Methods in Molecular Biology, vol. 2985, https://doi.org/10.1007/978-1-0716-4972-5_16,

genome architecture. Even more, comparative genomics approaches have been instrumental in uncovering core genomic elements, lineage-specific gene families, and structural variations that define the genetic architecture across different plant lineages [1–3].

A few bioinformatic methods, largely shaped by sequencing technologies, were applied to identify gene families linked to plant traits across lineages [1–3]. However, most bioinformatics pipelines were applied to plant genomes that are phylogenetically closely related [4]. As expected, the divergence range has a great impact on comparative analysis, since the longer the divergence times, the more likely for the genomes to accumulate more genetic singularities that hinder this sort of comparative genomic studies.

So, there remains much to be explored when we aim to identify gene-controlling traits. A challenging step is defining orthologous genes or orthogroups across divergent plant lineages, which are genes that originated from a common ancestor. Several major strategies have been developed to identify orthologous genes. The first strategy involves phylogenetic tree analysis, based on comparing gene trees with species trees [5]. These methods distinguish orthologs as homologs that diverged through speciation events rather than gene duplication. Although conceptually ideal for orthology identification, phylogenetic trees are computationally consumptive and prone to errors, particularly over large evolutionary distances. A second strategy uses heuristic algorithms that focus on sequence conservation to identify probable orthologs as the closest homologous pairs or groups of genes across organisms. These methods are faster and more amenable to automation than phylogenetic approaches. Although these heuristic methods differ conceptually from tree-based methods, they tend to produce similar sets of orthologs, especially when considering shorter evolutionary distances. Synteny analysis represents a third strategy, which can also aid in the identification of orthologs [6]. Synteny is the preservation of colinear chromosomal blocks when comparing more than two sets of chromosomes. By examining the conserved genomic context of genes, researchers can infer orthologous relationships that might be overlooked by sequence similarity alone. Nowadays, these approaches are more commonly used in conjunction with the aforementioned methods, leading to hybrid methods that leverage the strengths of each strategy. A specific tool that exemplifies the use of a hybrid heuristic algorithm for ortholog identification is *OrthoFinder* [7], an efficient and scalable method that uses graph-based similarity searching approaches and genes among species trees to delineate orthologous and co-orthologous genes. *OrthoFinder* can handle large datasets and provides a user-friendly way to compare multiple genomes, making it a valuable resource in the field of comparative genomics.

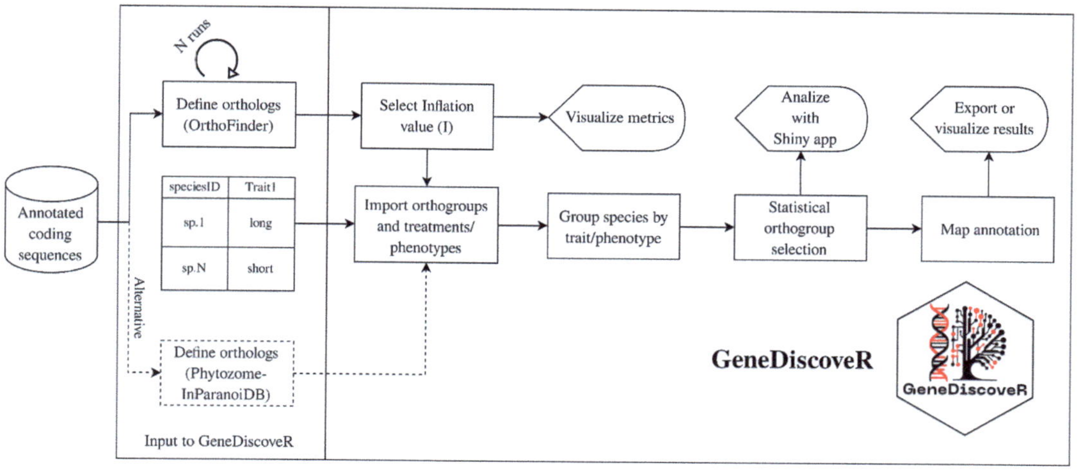

Fig. 1 Workflow of GeneDiscoveR toolbox for plant comparative genomics. The workflow encompasses two main parts: preparing inputs and using GeneDiscoveR. Input preparation with OrthoFinder is computationally intensive but allows the user to decide the optimal clustering and partitioning of orthogroups. In this regard, GeneDiscoveR facilitates testing different OrthoFinder inflation values to make a final selection. Alternatively, orthogroups can be imported from InParanoiDB files. The selected inflation value defines orthogroups and hierarchical orthogroups for further analysis. Users should prepare a file classifying species according to their traits, e.g., OB type, many in all cells or one in specialized cells, or self-incompatibility, or treatment response like salinity tolerance or pathogen susceptibility. GeneDiscoveR is a robust toolbox for selecting gene candidates by relating their presence or absence to a trait or treatment response. In this way, the trait will be associated with the enrichment of genes in the species that have it and absent in those that do not. The final output file includes an annotated file of orthogroups for each species and a Volcano plot depicting enrichment measure by odds ratio vs. statistical significance, enabling easy selection and output of candidate genes with their annotations

Therefore, it becomes imperative to develop robust bioinformatic protocols tailored to the unique evolutionary trajectories of land plants. Here, we introduce *GeneDiscoveR*, an R package designed for the statistical detection of orthogroups related to any plant trait facilitating the identification of specific orthogroups and genes that might be required for the trait. We demonstrate the process of identifying homologs and orthologs genes using extensive genomic data from diverse land plant lineages, exemplified here using liverwort genomic data for the identification of lineage trait-specific *TERPENE SYNTHASE* (*TPS*) genes and Brassicaceae genomes for the identification of self-incompatible related genes.

The bioinformatics pipeline outlined in this study provides general steps and parameters for analyzing large-scale genomic data, offering a roadmap for future gene discovery efforts in land plants. An overview of the workflow is provided in Fig. 1.

2 Materials

2.1 Annotated Coding Sequences and Identified Phenotypes per Species

Annotated coding sequences stored in multifasta files, one per species, are required for the identification of orthologous genes. If orthology has already been defined, with *OrthoFinder* or InParanoid, this can be the input for gene discovery. Additionally, a table with at least two columns indicating the species and the phenotype (s) to be analyzed is needed.

2.2 Bioinformatics Tools and Scripts

1. *Conda* (https://docs.anaconda.com/free/miniconda/index.html) is an open-source, cross-platform, language-agnostic package manager and environment management system (only required if orthology definition is missing).
2. *Snakemake* (https://snakemake.readthedocs.io/en/stable/) is a tool for creating reproducible and scalable data analyses. A Python-based language is used for creating workflows, which can be scaled to server, cluster, grid, and cloud environments without the need to modify the workflow definition (only required if orthology definition is missing).
3. *OrthoFinder* (https://github.com/davidemms/OrthoFinder) is a comparative genomics tool that allows defining orthogroups, inferring rooted gene trees for all orthogroups, and identifying all gene duplication events. Additionally, it enables the inference of a rooted species tree for the analyzed species.
4. *Python* with *numpy* (https://www.python.org/) and *R* (https://www.r-project.org/) are the programming languages used in the pipeline. Python code will be used to generate the input results for *GeneDiscoveR*.
5. *GeneDiscoveR* (https://github.com/AtilioRausch/GeneDiscoveR/) is an *R* package for the statistical detection of orthogroups related to different treatments or phenotypes associated with the presence or absence of genes. It can be executed in an *R* environment using the orthogroups defined.

It is advisable to execute the first step of this pipeline (defining orthologs with *OrthoFinder*) on a server with multiple cores for fast and scalable execution. However, the pipeline can be executed on any PC.

3 Methods

The workflow for identifying genes whose presence or absence is associated with a trait, treatment response, or phenotype is shown in Fig. 1. The workflow is divided into two major blocks. The first block prepares the input data for *GeneDiscoveR*, offering multiple options, such as Define orthologs (*OrthoFinder*) or from *InParanoiDB*, as can be obtained from the plant genomics portal Phytozome. The second block, using *GeneDiscoveR*, involves the following steps: Select the granularity of the similarity graph controlled by the inflation value (*I*) (*see* **Note 1**), import orthogroups and treatments/phenotypes, group species by trait/phenotype, perform statistical orthogroup selection, map annotation, and visualize or export results.

As examples of the pipeline application, we first demonstrated the identification of *TPS* genes associated with different types of oil bodies (OBs) in liverworts: some with OBs within specialized cells (idioblasts) and others with many OBs in all cells. We analyzed the annotated coding sequences of 42 liverworts (unpublished data), including 27 species with many OBs in all cells (Jungermanniopsida and Haplomitriales) and 10 species that form OBs in idioblasts (Marchantiopsida). Additionally, five species that are not capable of forming OBs were also included. Secondly, the pipeline was applied to analyze the phenotype associated with self-incompatibility based on the annotated coding sequences of 27 species from the Brassicaceae family (Phytozome), where 8 were identified as self-compatible and 19 as self-incompatible. The description of the analyzed species and their classification can be found in /inst/extdata/annotatedCDSs.tsv and /inst/extdata/Brassicaceae/table_traits_selfcomp.tsv for the OB-type and self-incompatibility phenotypes, respectively.

3.1 Define Orthologs (OrthoFinder)

Initially, the workflow starts in the shell prompt. As an initial step, groups of orthologous/homologous genes must be defined. For this, *OrthoFinder* is used, which is capable of defining orthogroups (*OGs*) and phylogenetic hierarchical orthogroups (*HOGs*) from the annotated coding sequences of multiple species. We recommend running *OrthoFinder* multiple times, varying the inflation value (*I*) parameter (*see* **Note 1**).

This is the most computationally costly step in the workflow, so it is recommended to run using multiple cores. To do this, initially create and activate a *Conda* environment with *OrthoFinder*, *SnakeMake*, and *Numpy*:

```
conda create -n env-ortho -c bioconda orthofinder snakemake
numpy
conda activate env-ortho #Activate environment
```

With the active environment, execute the following command:

```
snakemake --use-conda --cores 32 --snakefile run_orthofinder.
py --configfile config.yaml > log_orthofinder.log
```

where the arguments are `--use-conda` to use *OrthoFinder* from the active *Conda* environment and `--cores` as the number of cores to use. Additionally, the script to be executed (`--snake-file`) is `run_orthofinder.py` and the configuration file (`--configfile`) is `config.yaml`.

The `run_orthofinder.py` script and the `config.yaml` configuration files are located in /Python. Modify the config.yaml file indicating the directory where your annotated coding sequences files are located and the inflation values, by default, 48 runs, will be performed varying the inflation value from 1.3 to 6 with a step of 0.1.

As a result of the execution, 48 directories are obtained. From these directories, the N0.tsv files, which contain the *HOGs*, and the Statistics_Overall.tsv files for each execution must be extracted. To do this, modify the input ("path/to/results/Orthofinder/") and output paths of the `collect_rename_overall_statistics.sh` and `collect_rename_N0s_files.sh` scripts and execute:

```
./collect_rename_N0s_files.sh
./collect_rename_overall_statistics.sh
```

Both scripts are located in /bash.

In our example, since the definition of orthologous is the only execution step external to *GeneDiscoveR*, the results of the executions already extracted and renamed are found in /inst/extdata/Comparatives-1dot3-6 and /inst/extdata/N0-1dot3-6. On the other hand, if you do not have access to multiple cores, you can perform a run using the default inflation value of 1.5 as in the Brassicaceae example, or you can start the analysis using Phythozome's InParanoiDB files.

3.2 Install and Import GeneDiscoveR

We present *GeneDiscoveR*, an R package for the statistical detection of orthogroups related to different treatment responses or phenotypes associated with the presence or absence of genes. *GeneDiscoveR* is implemented in a modular way with S3 classes, which provides great versatility in the analysis of its results. In addition, it allows input from multiple ortholog definition tools and facilitates execution with multiple cores, making it scalable and fast. Additionally, it provides multiple ways to visualize and export the results, including interactive graphics and shiny apps.

1. *GeneDiscoveR* is installed from its GitHub repository https://github.com/AtilioRausch/GeneDiscoveR. In an R environment, execute

```
library("usethis")
library("devtools")

        devtools::install_github("AtilioRausch/GeneDiscoveR")
```

2. *Load GeneDiscoveR* package

```
library(GeneDiscoveR)
```

3.3 Import Orthogroups and Traits/Phenotypes

As inputs, *GeneDiscoveR* receives the paths of the two directories resulting from Subheading 3.1, which contain the Statistics_Overall.tsv and N0.tsv files for each run. In addition, the path of a tsv file with at least two columns must be indicated, one for the species *IDs* used by *OrthoFinder* and one or more columns for the *n* phenotypes to be analyzed (*see* **Note 2**).

1. Define the input paths:

```
overallsDir <- "/path/to/dir/with/Statistics_OverallFiles/"
#From Section 3.1N0sDir <- "/path/to/dir/with/N0files/" #From
Section 3.1dataFile <- "/path/to/file/with/treatments.tsv"
```

In this example, we use

```
overallsDir <- system.file("extdata", "Comparatives-1dot3-6",
package = "GeneDiscoveR")
N0sDir <- system.file("extdata", "N0-1dot3-6", package =
"GeneDiscoveR")
dataFile <- system.file("extdata", "annotatedCDSs.tsv", pack-
age = "GeneDiscoveR")
```

2. Create a *GeneDiscoveR* object indicating the paths and the minimum, maximum, inflation values and the step used for the screening:

```
GeneDiscoveRobject <- GeneDiscoveR(
 overallsDir = overallsDir,
 N0sDir = N0sDir,
 dataFile = dataFile,
 minInflation = 1.3,
 maxInflation = 6,
 stepInflation = 0.1
)
```

If you run *OrthoFinder* for a single inflation value, you can indicate the inflation value used with the `uniqueInflation` parameter.

As a result, a *GeneDiscoveR* object is obtained, containing the necessary information to be analyzed using the following functions.

3.4 Select Inflation Value (I)

To select the inflation value, the balance between clustering and partitioning must be considered. Excessive clustering can result in an orthogroup containing multiple gene families, whereas excessive partitioning may cause genes from the same family to be left out of the orthogroup. In addition, the root clade inferred by *STRIDE* [8] should be considered.

1. First, obtain the statistics for each run, highlighting the number of orthogroups that contain at least one gene from all species. This is important as the *STAG* [9] algorithm needs at least 100 orthogroups with genes from all species to obtain a root species tree.

```
GeneDiscoveRobject <- calculate_overall_statistics(GeneDis-
coveRobject, cores = 8)
```

2. Second, to analyze the clustering, get the number of *HOGs* and *OGs* per run. Here, metrics associated with the partition in *HOGs* for each *OG* are also calculated.

```
GeneDiscoveRobject <- calculate_median_statistics(GeneDis-
coveRobject, cores = 8)
```

3. To access the data, you can use the `get_overall_statistics()` and `get_medians_statistics()` functions, respectively. You can also visualize the results in the following way:

```
plot_OGs_HOGs_per_inflation(GeneDiscoveRobject) #see Note 3
and Fig. 2A
plot_allSpeciesOGs_sOGs_per_inflation(GeneDiscoveRobject)
#see Note 3 and Fig. 2B
```

4. In this example, the value of $I = 1.8$ represents a midpoint between clustering and partitioning (*see* **Note 3**). It is also the last value of I for which the *Haplomitrium-Treubia* clade in our liverwort example was inferred as the root, and more than 100 *OGs* are defined with at least 1 copy of the gene for each species (*see* **Note 3**). The clade inferred as the root for the species tree by *STAG* can be observed in the log_orthofinder.log file obtained in the multiple executions of *OrthoFinder* under the subtitle "Best outgroup(s) for species tree" for each execution.

   Set active execution:

```
GeneDiscoveRobject <- set_run_active(GeneDiscoveRobject, In-
flationValue = 1.8)
```

This step is optional but recommended. You can skip it if you performed a single *OrthoFinder* execution or if you are using *inParanoiDB* files from the plant genome database *Phytozome* or another source.

3.5 Group Species by Trait/Phenotype

Group the species by phenotype or trait, indicating in each case the column name of the phenotype to analyze "Oil-body-type," the column where the *ID* of the species "OrthofinderID" is found, and the type in each case "one_in_specialized" and "many_in_all_cells" as corresponds in annotatedCDSs.tsv.

1. Group species with OB within specialized cells:

```
GeneDiscoveRobject <- select_species_by_phenotype(
 GeneDiscoveRobject = GeneDiscoveRobject,
 columnPhenotype = "Oil-body-type",
 columnID = "OrthofinderID",
 type = "one_in_specialized_cell"
)
```

2. Group species with many OB in all cells:

```
GeneDiscoveRobject <- select_species_by_phenotype(
 GeneDiscoveRobject = GeneDiscoveRobject,
 columnPhenotype = "Oil-body-type",
 columnID = "OrthofinderID",
 type = "many_in_all_cells"
)
```

3.6 Statistical Orthogroup Selection

1. To distinguish the *OGs* that are enriched with genes from species with one phenotype or another, Fisher's exact test is applied:

```
GeneDiscoveRobject <- gene_identification_by_phenotype(
 GeneDiscoveRobject = GeneDiscoveRobject,
  formula = as.formula("many_in_all_cells ~ one_in_speciali-
zed_cell"),
 statistic = "Fisher",
 name = "PerType",
 cores = 8
)
```

In this case, the function applied Fisher's test to identify the *OGs* significantly enriched in genes from species with the "many_in_all_cells" phenotype ($p - value < 0.05$ *and Odds ratio* > 1) and absent in the species with the "one_in_specialized_cell" phenotype and the *OGs* enriched in genes from the species with the "one_in_specialized_cell" phenotype and absent in the species with the "many_in_all_cells" phenotype ($p - value < 0.05$ *and Odds ratio* < 1).

Multiple identifications can be performed on the same *GeneDiscoveR* object with different formulas and considering multiple phenotypes. It is important to rename the identification for proper storage of the identifications.

2. To obtain the *OGs* significantly enriched in genes from species with OB in specialized cells ($p - value < 0.05$ *and Odds ratio* > 1), execute

```
GeneDiscoveRobject <- select_genes_by_phenotype(GeneDiscoveR-
object,
 pvalue = 0.05,
 oddsRatio = 1,
 sign = ">",
 name = "PerType"
)
```

The filtered table will be stored in your *GeneDiscoveR* object. You can make multiple selections and access the resulting tables with the function `get_filtered_genes_table ()`.

3.7 Map Annotation

1. To identify each orthogroup, it is interesting to map the functional annotation. For this purpose, the annotation from a model species can be used. For the annotation of genes from non-model species, *see* **Note 4**. In our example, the annotation from *Marchantia polymorpha* subsp. rudelaris is imported from the MarpolBase database:

```
annotationFile <- system.file("extdata", "MpTak_v6.1_func_an-
notation_1line.tsv",
 package = "GeneDiscoveR")
GeneDiscoveRobject <- set_annotation_file(GeneDiscoveRobject,
 annotationFile = annotationFile)
```

The annotation file should have the gene IDs in the first column and the functional annotation in one or more columns. If the annotation is in a single column, the separator character between fields must be identified.

2. To map the functional annotation to the filtered table, you need to identify the selection made in Subheading 3.6, **step 2**:

```
indexFilteredGenes <- select_filtered_gene_index(GeneDiscov-
eRobject,
 name = "PerType", p-value = 0.05,
 oddsRatio = 1, sign = ">")
```

3. Map the functional annotation:

```
GeneDiscoveRobject <- map_annotation(GeneDiscoveRobject =
GeneDiscoveRobject,
 indexFilteredGenes = indexFilteredGenes,
  specieWithAnnotation = "MpTAKv6-March-
antia_polymorpha_rudelaris",
 oneColumn = TRUE, sep = ";"
)
```

In this case, `map_annotation()` is used, specifying the ID of the species with annotation. Since the functional annotation fields are in a single column, `oneColumn=TRUE` is used, and the separator is specified as `sep=";"`. On the other hand, if you want to map the functional annotation to all *OGs*, you should specify `indexFilteredGenes=NULL`.

3.8 Visualize and Export Results

GeneDiscoveR provides flexibility for storing and exporting the data. If you want to export a table of filtered genes with their annotation, you can use `get_filtered_genes_table()`. However, if you want to obtain the complete table of orthogroups, you can use `get_complete_table()`.

When it comes to visualization, you can run a shiny web app for analysis, visualizations, and data export with `run_genediscov-er_web_app()`. In addition, you can obtain the names of the identifications performed with `get_names_identification ()`. Then, to visualize the Volcano plot, execute the function `plot_genediscover_volcano()`.

1. In this context, you can identify the OGs to which a list of genes belongs and visualize them in a Volcano plot. To do this, you must initially obtain the OGs in which your genes are contained. In this example, we start with a table of identified Terpene-Synthase genes (*see* **Note 5**).

```
TPSgenes <- read_tsv(system.file("extdata", "TPSgenes.tsv"),
col_names = T)
```

2. Obtain the list of genes and species identified by their OrthoFinder ID:

```
GeneID <- TPSgenes$GeneID[TPSgenes$TPStype == "Bacterial"]
OrthoFinderID <- TPSgenes$OrthofinderID[TPSgenes$TPStype ==
"Bacterial"]
```

3. To obtain the OGs to which your genes belong, execute

```
Bacterial <- obtain_OG_from_gene(GeneDiscoveRobject, GeneID,
OrthoFinderID)
```

The resulting table contains the columns GeneID, OG, and Gene Tree Parent Clade.

4. Finally, add a new column that defines the category, in this case, TPStype:

```
Bacterial <- Bacterial %>% mutate(TPStype = "Bacterial")
```

It is important to note that information about bacterial *TPS* genes is extracted in the steps performed. If you have more than one classification for the same category, repeat the steps while maintaining the name of the categorical column, e.g., TPStype in **step 4**, and concatenate the tables with `TPSs <- bind_rows(Bacterial, Diterpeno, Fungi)`.

5. After obtaining the OGs, the user can visualize the Volcano plot (Fig. 3), which indicates the categories (`TPScategories`), the table obtained from Subheading 3.8, **step 4** (`TPSs`), the orthogroup identification name from Subheading 3.6, **step 1** (`name = "PerType"`), and the name of the categorical variable from **step 4** (`type = "TPStype"`).

```
TPScategories <- c("Diterpene", "Bacterial", "Fungi")
title <- "Fisher's Exact Test per Type of Oil-body with TPS
genes"
plot_genediscover_detector_volcano(GeneDiscoveRobject,
 annotationTable = TPSs, title = title,
 name = "PerType", type = "TPStype",
 categories = TPScategories
)
```

The Volcano plot combines statistical significance (represented by the p-value) and the strength of association (represented by the odds ratio) to highlight differentially enriched OG types in the context of OB distribution. Additionally, to demonstrate the potential of *GeneDiscoveR*, we include an overlaid layer of information mapping the *TPS* gene types onto the Volcano plot. Users will be able to submit their own set of genes. The full explanation for mapping genes onto the Volcano plot is described on the GeneDiscoveR GitHub examples and vignettes: https://github.com/AtilioRausch/GeneDiscoveR/.

3.8.1 Axes Explanation

- log(Odds Ratio): This axis shows the log-transformed odds ratio, which measures the enrichment of OGs in species with specific types of OB. A positive log(odds ratio) indicates enrichment in species with idioblasts, while a negative value indicates enrichment in species with many OB in all cells.

- -log(*p*-value): This axis represents the negative log-transformed *p*-value from the Fisher's exact test. Above the dashed red line, values correspond to statistical significance ($p - value < 0.05$).

3.8.2 Interpreting Enrichment

- The right side of the plot: Points on the right side (positive log (odds ratio)) indicate enrichment of OGs in species with OB in specialized cells. For example, orange dots (bacterial Mp*TPS* genes) and blue points (*diterpene* Mp*TPS* genes, *MpDTPS*) on the right indicate these gene types are more common in species with specialized OB.
- Left side of the plot: Points on the left side (negative log(odds ratio)) indicate enrichment of OGs in species with many OB in all cells. For instance, blue points (*MpDTPS* genes) on the left suggest these gene types are more common in species with widespread OB.

3.9 Using GeneDiscoveR for Different Traits in the Brassicaceae Family

To show the applicability of the package to evolutionary closer species and expand its use to a different plant lineage, we analyzed plants with sequenced genomes from the Brassicaceae family. This family presents several species of interest, such as the well-studied model plant *Arabidopsis thaliana* and crops like *Brassica napus* and *Brassica rapa*. These species exhibit diverse traits that can be explored for gene discovery, including salt tolerance, metal hyper-accumulation, and different life histories [10]. One extensively studied trait in this family is the barrier to self-pollination and the related trait selfing syndrome, an umbrella term describing the typical changes that the flower suffers when a species becomes self-compatible [11]. In contrast, plants unable to self-pollinate, also called outcrossers, present self-incompatibility. We chose this trait for the *GeneDiscoveR* analysis.

We first performed the orthology definition with *OrthoFinder* on the annotated coding sequences files of 27 Brassicaceae species, retrieved from the Phytozome database [12], using default inflation parameter. Then, we run the *GeneDiscoveR* analysis pipeline. As a result, we identified 66 orthogroups significantly associated with the self-incompatibility trait, among which OG0005891 stands out (Fig. 4). This orthogroup is composed of genes from 16 out of 19 self-incompatible species and 3 out of 8 self-compatible species. The genes belonging to the orthogroup OG0005891 exhibit the Plant self-incompatibility S1 domain (Fig. 4), known to possess specific pollen-inhibitory properties [13, 14].

Our results demonstrate that *GeneDiscoveR* can be utilized to identify novel candidate genes within the Brassicaceae family, some of which have not yet been described, even in the model organism *Arabidopsis thaliana*.

4 Notes

1. *OrthoFinder* facilitates the identification of orthogroups (*OGs*) and phylogenetic hierarchical orthogroups (*HOGs*). *OGs* are obtained through all-vs-all sequence comparisons and the creation of a graph based on sequence similarity. In contrast, *HOGs* are derived from the analysis of gene trees rooted with the clade root inferred by the *STRIDE* algorithm [8]. This methodology offers 12% greater accuracy compared to the traditional gene similarity graph approach. In this context, the inflation value (I) controls the clustering of the Markov Cluster Algorithm [15] (*MCL*) applied to the graph of similar genes. If the inflation value is too high, there is a risk of excluding several genes from the same family from the *OG*. Conversely, if it is too low, it may result in excessive clustering that includes multiple gene families. In the latter case, the correct definition of the *OG* will depend exclusively on the hierarchical partitioning into *HOGs*, which is contingent on the accuracy of the species tree.
2. As input files, in all cases, *GeneDiscoveR* requires a table with at least two columns: one with the species ID used by the ortholog definition tool and one or more columns for the phenotypes to be studied. For example, Table 1 shows the ID used by *OrthoFinder* and two phenotypes, the presence of oil body and type of oil body. Additionally, if you performed a run with multiple inflation values, you must indicate the path to the directories containing the N0s files and Overall Statistics files (optional). If you performed a single run of OrthoFinder, indicate the path to the N0 file. Lastly, if you obtained inParanoiDB-Phytozome files, you must indicate the path to the directory with the files that relate your main species to the rest of the species.
3. The execution of *OrthoFinder* with an inflation value of 1.8 ($I = 1.8$) represents an intermediate point between clustering and partitioning (Fig. 2a). Additionally, it is the last I value for which the *Haplomitrium-Treubia* clade is inferred as the root, and more than 100 *OGs* are defined with at least 1 copy per species (Fig. 2b).
4. To annotate genes from non-model species belonging to a specific orthogroup, you can start by identifying the orthogroup using the `obtain_OG_from_gene()` function, retrieve a table of genes that meet certain conditions with `get_filtered_genes_table()`, or obtain the complete orthogroup table with `get_complete_table()`. Once the orthogroup of interest is identified, for example, OG0005397, you can obtain the gene sequences from /path/to/results/orthofinder/Orthogroup_Sequences/OG0005397.fa.

Table 1
Example of input table with N species and M traits or treatment response, one per column

OrthofinderID	Oil-body-presence	Oil-body-type	
1016-*Treubia_lacunosa*	1	one_in_specialized_cell	M traits or treatment response
M158-*Sphaerocarpos_donnellii*	0	noOB	
225-*Ptilidium_ciliare*	1	many_in_all_cells	
1017-*Porella_caespitans*_var_nipponica	1	many_in_all_cells	
N species			

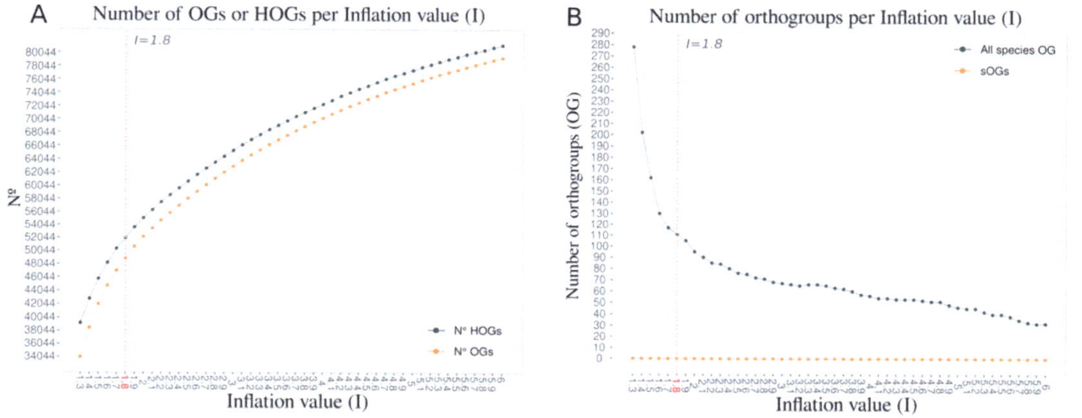

Fig. 2 Selection of inflation value using *OrthoFinder*. (**a**) Number of *OGs* and *HOGs* per inflation value. (**b**) Number of *OGs* and single copy *OGs* per inflation value with genes from all species. A vertical dashed line indicates the chosen inflation value using `set_run_active()`. The figures are generated using the functions `plot_OGs_HOGs_per_inflation()` and `plot_allSpeciesOGs_sOGs_per_-inflation()` after selecting the inflation value

Then, outside the R environment, you need to create a Conda environment and install HMMER [16]:

```
conda create -n hmmer -c bioconda hmmer
conda activate hmmer
```

Download the Pfam [17] domains database from https://ftp.ebi.ac.uk/pub/databases/Pfam/releases/[version]/Pfam-A.hmm.gz; replace `[version]` for the last Pfam version available, i.e. Pfam37.0; and run

```
hmmscan --domtblout OG0005397-annotation.tsv Pfam-A.hmm.gz
OG0005397.fa.
```

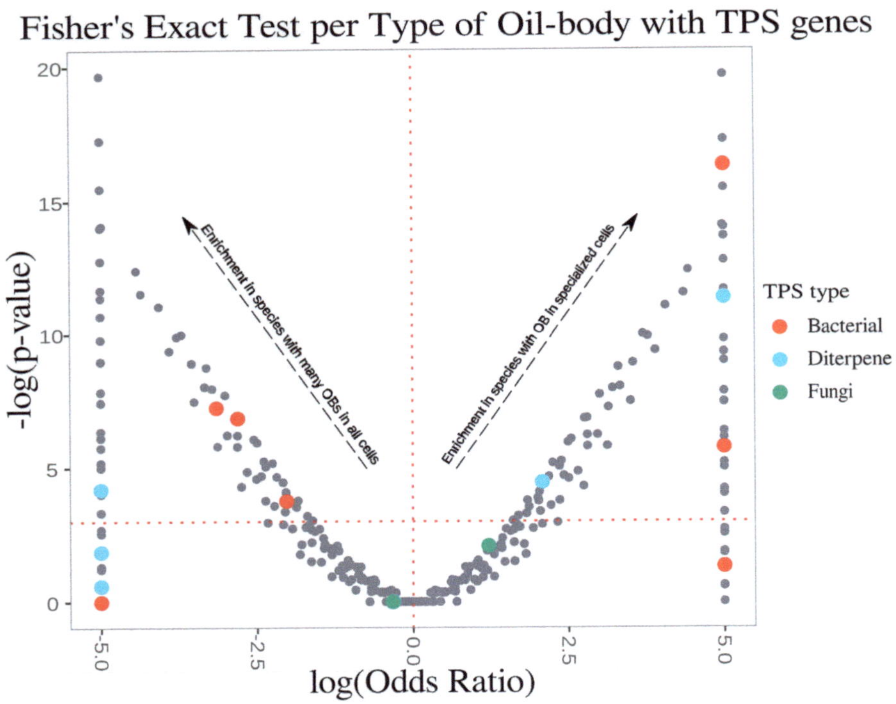

Fig. 3 Volcano plot generated using GeneDiscoveR. This volcano plot provides a visual summary of the association between OGs and OB distribution in different species. In the particular case of TPS genes, by examining the position and color of the points, users can identify which TPS gene types are significantly enriched in species with different OB characteristics, aiding in understanding the functional roles of these genes in plant biochemistry and evolution

The resulting OG0005397-annotation.tsv file will contain a table listing the domains present in the genes of orthogroup OG0005397.

5. *TPS* genes were identified from the annotated coding sequences of 38 liverwort species, including *M. polymorpha* subs. rudelaris. For this, *HMMER* [16] (v3.3.2) was used with conserved profiles for the C-terminal domain (PfamID: PF03936), C-terminal metal binding (PfamID: PF19086), N-terminal (PfamID: PF01397), and trichodiene synthase (PfamID: PF06330) with an e-value < 1e-5, identifying 502 sequences. It is important to note that the microbial *TPS* (*MpMTPS)* and diterpene synthase (*MpDTPS*) previously identified in *M. polymorpha* were added [18, 19]. An alignment was then performed using *Muscle* (v5.1) [20], followed by two filtering steps: with TrimAL (v1.4rev15) [21] using the `-gappyout` option and with *BMGE* (v1.12) [22] without trimming based on entropy and filtering sequences and columns with than 40% and 80% gaps, respectively. Finally, *IQ-TREE* (v2.0.3) [23] was used to create a maximum likelihood tree

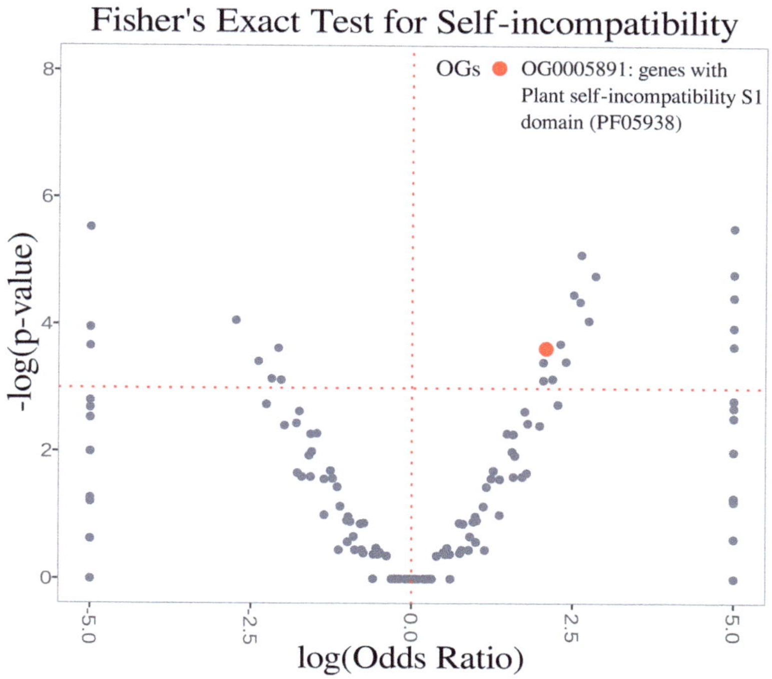

Fig. 4 Volcano plot obtained with the function plot_genediscover_detector_volcano() from GeneDiscoveR for identifying orthogroups associated with the self-incompatibility phenotype in species of the Brassicaceae family. The red circle highlights orthogroup OG0005891, which contains the gene AL3G21840 from *Arabidopsis lyrata*, known to have the Plant self-incompatibility S1 domain

with 1000 bootstrap iterations and automatic detection of the evolutionary model. The tree branches were classified based on the position of previously identified *MpMTPS* (*MpMTPSL6*, *MpMTPSL7*, and *MpMTPSL9* from fungi; *MpMTPSL1*, *MpMTPSL2*, *MpMTPSL3*, *MpMTPSL4*, *MpMTPSL5*, and *MpMTPSL8* from bacteria) and *MpDTPS* (*MpDTPS1*, *MpDTPS2*, *MpDTPS3*, and *MpDTPS4* from plants) genes as diterpene synthase from plants.

Acknowledgments

This work was supported by research grants from Agencia I + D + i (PICT2018-0650, PICT2020-0110 to JEM) and CONICET Research Program to JEM and Scientific Foundation of the Urban Management Bureau of Shenzhen (202403 to YL). PhD fellowships from CONICET and Agencia I + D + i P support the PhD studies of AOR, JIT, and TMR.

Conflict of Interest The authors have no conflicts of interest to declare that are relevant to the content of this chapter.

References

1. Tello-Ruiz MK, Naithani S, Gupta P et al (2021) Gramene 2021: harnessing the power of comparative genomics and pathways for plant research. Nucleic Acids Res 49:D1452–D1463. https://doi.org/10.1093/nar/gkaa979
2. Kress WJ, Soltis DE, Kersey PJ et al (2022) Green plant genomes: what we know in an era of rapidly expanding opportunities. Proc Natl Acad Sci USA 119. https://doi.org/10.1073/pnas.2115640118
3. Pucker B (2022) Automatic identification and annotation of MYB gene family members in plants. BMC Genomics 23:220. https://doi.org/10.1186/s12864-022-08452-5
4. Xu Z, Sun W, Zhu Z et al (2023) Phylogenetic inference of homologous/orthologous genes among distantly related plants. Bio-protocol 13:e4893. https://doi.org/10.21769/BioProtoc.4893
5. Szöllősi GJ, Tannier E, Daubin V, Boussau B (2015) The inference of gene trees with species trees. Syst Biol 64:e42–e62. https://doi.org/10.1093/sysbio/syu048
6. Almeida-Silva F, Zhao T, Ullrich KK et al (2023) Syntenet: an R/Bioconductor package for the inference and analysis of synteny networks. Bioinformatics 39:btac806. https://doi.org/10.1093/bioinformatics/btac806
7. Emms DM, Kelly S (2019) OrthoFinder: phylogenetic orthology inference for comparative genomics. Genome Biol 20:238. https://doi.org/10.1186/s13059-019-1832-y
8. Emms DM, Kelly S (2017) STRIDE: species tree root inference from gene duplication events. Mol Biol Evol 34:3267–3278. https://doi.org/10.1093/molbev/msx259
9. Emms DM, Kelly S (2018) STAG: species tree inference from all genes. bioRxiv. https://doi.org/10.1101/267914
10. Nikolov LA, Tsiantis M (2017) Using mustard genomes to explore the genetic basis of evolutionary change. Curr Opin Plant Biol 36:119–128. https://doi.org/10.1016/j.pbi.2017.02.005
11. Tsuchimatsu T, Fujii S (2022) The selfing syndrome and beyond: diverse evolutionary consequences of mating system transitions in plants. Philos Trans R Soc London Ser B Biol Sci 377:20200510. https://doi.org/10.1098/rstb.2020.0510
12. Goodstein DM, Shu S, Howson R et al (2012) Phytozome: a comparative platform for green plant genomics. Nucleic Acids Res 40:D1178–D1186. https://doi.org/10.1093/nar/gkr944
13. Nasrallah JB, Nasrallah ME (2014) S-locus receptor kinase signalling. Biochem Soc Trans 42:313–319. https://doi.org/10.1042/BST20130222
14. Ivanov R, Fobis-Loisy I, Gaude T (2010) When no means no: guide to Brassicaceae self-incompatibility. Trends Plant Sci 15:387–394. https://doi.org/10.1016/j.tplants.2010.04.010
15. Van Dongen S (2008) Graph clustering via a discrete uncoupling process. SIAM J Matrix Anal Appl 30:121–141. https://doi.org/10.1137/040608635
16. Finn RD, Clements J, Arndt W et al (2015) HMMER web server: 2015 update. Nucleic Acids Res 43:W30–W38. https://doi.org/10.1093/nar/gkv397
17. Mistry J, Chuguransky S, Williams L et al (2021) Pfam: the protein families database in 2021. Nucleic Acids Res 49:D412–D419. https://doi.org/10.1093/nar/gkaa913
18. Kumar S, Kempinski C, Zhuang X et al (2016) Molecular diversity of terpene synthases in the liverwort Marchantia polymorpha. Plant Cell 28:2632–2650. https://doi.org/10.1105/tpc.16.00062
19. Jia Q, Li G, Köllner TG et al (2016) Microbial-type terpene synthase genes occur widely in nonseed land plants, but not in seed plants. Proc Natl Acad Sci USA 113:12328–12333. https://doi.org/10.1073/pnas.1607973113
20. Edgar RC (2004) MUSCLE: multiple sequence alignment with high accuracy and high throughput. Nucleic Acids Res 32:1792–1797. https://doi.org/10.1093/nar/gkh340
21. Capella-Gutiérrez S, Silla-Martínez JM, Gabaldón T (2009) trimAl: a tool for automated alignment trimming in large-scale phylogenetic analyses. Bioinformatics 25:1972–1973. https://doi.org/10.1093/bioinformatics/btp348

22. Criscuolo A, Gribaldo S (2010) BMGE (block mapping and gathering with entropy): a new software for selection of phylogenetic informative regions from multiple sequence alignments. BMC Evol Biol 10:210. https://doi.org/10.1186/1471-2148-10-210

23. Minh BQ, Schmidt HA, Chernomor O et al (2020) IQ-TREE 2: new models and efficient methods for phylogenetic inference in the genomic era. Mol Biol Evol 37:1530–1534. https://doi.org/10.1093/molbev/msaa015

Index

José M. Franco-Zorrilla and Carmen Martín-Pizarro (eds.), *Plant Transcription Factors: Methods and Protocols*, Methods in Molecular Biology, vol. 2985, https://doi.org/10.1007/978-1-0716-4972-5,

R

S

T

W

Y